THE JOURNAL OF BEAVER ISLAND HISTORY

Volume Two

Essays on the history of Beaver Island

In this series:
THE JOURNAL OF BEAVER ISLAND HISTORY Volume I

Compiled, edited and distributed by the
Beaver Island Historical Society,
St. James, Michigan 49782

First printing July 1980

Printed by
Wellman Press, Inc.
200 N. Cedar St.
Lansing, Michigan 48901

Preface

The Beaver Island Historical Society is happy to bring forth THE JOURNAL OF BEAVER ISLAND HISTORY, Volume II, This is the second in a series subtitled Essays on the History of Beaver Island, intended to help preserve and interpret the fascinating events and lore which have affected the island and its people.

The first volume was published in 1976 by the Beaver Island Bicentennial Committee as a Bicentennial project and funded by a grant from the Michigan American Revolution Bicentennial Committee. When that year of celebration ended, the local committee turned over all of its projects to the Beaver Island Historical Society, so the 1980 volume is being published and funded by our society which is now celebrating its twenty-fifth year of activity.

The Historical Society has had eventful times since the Bicentennial year.

The projected restoration of the Old Post Office adjoining the Mormon Print Shop, which was mentioned as a goal in the Preface to Volume I, has been fully accomplished. It provides the building, which is used as the society's museum, with office space and a fine addition to our exhibit area. The remainder of the building has undergone some renovations and the collections are being organized and catalogued, so the museum is beginning to present an attractive new face to the public.

The Marine Museum has just been opened to visitors. It began as a Bicentennial project, and much but gradual work has turned the old net shed into a display and exhibit area intended to preserve and explain the history of the harbor area. It seems an auspicious beginning, and the public is already enjoying the results of the society's work.

The volunteer hours that have gone into these projects are countless! Many, many people have helped to achieve the results we now see. But this is the special time to thank the editorial board for their work on this volume: Barbara Beckers has served as Chairperson for the book and has been assisted by Phyllis Townsend and Mary Ann Omer. The rest of the board members and I wish to express our gratitude to them for giving of their talents and time, and trust that they will be willing to continue with work on Volume III! Plans and materials for the next in the JOURNAL OF BEAVER

ISLAND HISTORY series are already underway, so if you have stories, articles, ideas or suggestions for the next book, please contact us or the editorial board.

Gwin Langford, President BIHS

Beaver Island Marine Museum

Contributors

HELEN COLLAR, who contributed to Volume I as well, is a long-time summer resident and a lady of many talents. Her portraits of early Irish settlers hang in our museum, and her research and articles have established her as an authority about the island's past. We hope to see the culmination of a life's work published in her own volume soon.

BROTHER EDWARD FALLON has often visited "The Brothers' Place" and is on the faculty of Saint Patrick's High School in Chicago.

DOYLE FITZPATRICK is retired and lives on Beaver Island. He has long had an interest in the island's history and is well known as the author of THE KING STRANG STORY and as an authority on Strang history.

PHIL GREGG contributed to Volume I and was, for many years, the editor of "The Beaver Beacon" where many Beaver Tales have appeared.

MARGARET HANLEY again lets us have a view of island life from the days when she served as reporter of its doings.

GLEN HENDRIX'S studies in music and his concern and interest in the Beaver Island variety have grown with many summers spent here.

GRACE HILL and her husband, Walker, have both recently retired from Michigan State University and regularly spend the summers on Beaver.

MATTHEW HOHN is a member of the faculty at Central Michigan University and has been the Coordinator for the Biological Station since its beginning.

KEEWAYDINOQUAY has great concern for preserving the lore and traditions of Native Americans for their continuing use. She has spent many summers on Garden Island and is engaged in botanical research as it pertains to Indian culture.

AL MONTAGUE also retired to the island and has focussed his thirty-year interest in ships on our local history.

ANTJE PRICE, an occupational therapist from Pittsburgh, is known in St. James as "The Protar Lady." We hope to see her contributions to both Volume I and II incorporated with her continuing research on Protar become a book in its own right soon.

Mormon Print Shop before and after restoration.

Table of Contents

The Steamer "Eugene Hart" at dock.

BEAVER ISLAND AND THE BIG BOATS

by Al Montague

Before the income gained by catering to the needs of tourists and vacationers became a major factor in the Beaver Island economy there were four principal ways in which Islanders obtained their livelihoods. There were fishing and lumbering, occupations that were well covered in illustrated articles by Bill Cashman and Dave Gladish in *The Journal of Beaver Island History Volume I.* There was farming, which was usually on a subsistence basis without significant export for cash. Then there was the service of Beaver Island men on the large lake vessels whose sails and smoking stacks constantly marked the Island horizons. Men usually shipped out for nine months at a time on the "long ships," the ore boats and other bulk carriers that took cargos of coal, limestone, grain and ore to ports on the Great Lakes. They served first on the schooners and then on the old steam barges, "lumber hookers," some of which came to Beaver Island to pick up their cargos during logging days. Many Islanders served in the unique Lake Michigan car ferry fleets and also on the small tankers which serviced lake cities year around. In the heyday of lake passenger steamers that ended in the 1930's there were also many Islanders among the crews of those often luxurious vessels.

It is hardly remarkable that the inhabitants of an island took to the water and became skilled in the navigation of it. Island boys grew up seeing vessels, smaller and more numerous in times past than now, passing the Island on the northwest, past Lansing Shoals and on the east taking the Gray's Reef passage. Boats took refuge and still do from westerly gales in Sand Bay and from southerly winds to the north of Garden Island. It is only natural that the boys saw the distant ships first as an opportunity for travel and adventure and then as a means of earning money to send home to aging parents and growing families. In earlier times pay was low, living conditions rough and, in the days before radio, radar and modern aids to navigation, dangerous as the remains of many wrecks littering the bottoms and shores of the Great Lakes will verify. Until recent years when the right to regular leaves during the sailing season was gained, men were away from home for nine months at a time, often without stepping off the boat. Nevertheless men flocked to the big boats and, as Great Lakes' historian Dwight Boyer remarked in his book *Ghost Ships of the Great Lakes,* "every able-bodied Beaver Island man that wasn't fishing took to the lake fleets."

Today, in 1980, there are few Island men of middle years or older who have not put in at least a few seasons sailing on the big boats. It might be said here, that in the language of the lakes, a vessel can be a thousand feet long and still be called a "boat" whereas a salt water sailor would scorn to call a craft a quarter that size anything but a "ship."

Beaver Island, in the very early times of the white man's coming, served as an important fueling point and port of refuge for lake vessels. James Strang's newspaper the *Northern Islander,* mentions 3000 cords of hardwood for sale at Galilee, at the southern end of the Island for steamboat fuel. Competition was apparently keen. The going price was a dollar fifty a cord. A "rat" offered it at a dollar twenty-five according to the *Northern Islander.*

Strang recorded that as many as a dozen steamers arrived weekly at Saint James or Galilee. Presumably there were also many sailing craft that touched these ports. It was remarked that in the severe winter of 1855-56 the first steamer did not reach Saint James until April 29th.

There were two vessels making regular runs out of Beaver Island in Mormon times. The schooner "Dolphin" made a weekly run to Mackinac Island with mail and passengers. The fare was one dollar. A packet sailed regularly to Grand and Little Traverse Bays.

In earlier times shipwrecks were frequent among the rocks and shoals surrounding the Beaver Island group. Fog, ice and fall gales took a heavy toll, particularly among the thousands of schooners that sailed the lakes in the second half of the nineteenth century. These, often cheaply built craft, were many times considered to be expendable by their owners once they had earned a reasonable profit. Schooner timbers littered the shores of the lakes including those of Beaver Island. Later the wooden steam barges, "lumber hookers," took over much of the schooners' trade. Both types of craft called at the Island during the lumbering days to take out brimming cargos including heavy deck loads of boards to build the houses of the growing lake cities. A typical steam barge was the "Niko" wrecked north of Garden Island in November of 1924. Its remains have made a popular exploration site for skin divers in the last several years. Another of this type of vessel that took lumber out of Saint James was the "Robert C. Wente" which was lost on Lake Michigan in July of 1911.

Strang often mentions shipwrecks of the 1850's in the *Northern Islander.* One was right in Saint James harbor. The propellor "Ohio" broke loose from McKinley's wharf in "the worst east wind in thirty years" and went ashore. The propellor "Illinois" was wrecked in the Fox Islands. The "Sovereign Of The Seas" and the "Patchin" both went down on Skilligallee. These wrecks were recorded in just the few years of publication of the *Northern Islander.* Other typical snipwrecks occurring in the Beaver Island area after the Mormon times were: The schooner "Chandler P. Wells" in November 1884 on Whiskey Island reef; the steamer "Lady Washington" off Shishiwah in September 1889; the schooner "Milwaukee Belle" south of Beaver Island in November 1886; the steamer "Panther" six miles northeast of Beaver Island in November 1910; the schooner "LaFreniere" on Hog Island reef in November 1886 and the schooner "Princess Palms" in the same area. The steamer "Eugene Hart" was another ship which, after leaving Beaver Island with passengers and freight, bound for Chicago, was never heard from again. Years later the stranding of the gasoline tanker "Oswald J. Boyd" on Simmons reef in November 1936 precipitated the famous "Marold" tragedy which was recounted in Volume I of the Beaver Island history.

Shipwrecks have always had a fascination for spectators if not for the participants. There was the opportunity of yarning endlessly about them beside the hot stoves in the long winters. There were sometimes efforts to rescue the freezing survivors and more important, especially in hard times, there was opportunity for the salvage of wrecked vessels and their cargos. Many a house, in lake cities, was built from the cast-up deck loads of wrecked lumber schooners.

Beaver Islanders over the age of twenty-five, or so, will well remember one of the four major lake vessels to be lost by storm in the last forty years. Just eighteen miles west of the Island near Boulder Reef, the Bradley Fleet self-unloader "Carl D. Bradley," one of the largest boats on the lakes, broke in two in the southwester of November 18, 1958 and plunged to the bottom in 400 feet of water. There were only two survivors out of her crew of thirty-five. Many Islanders went to the western bluffs of the Island to watch for the lights of possible life rafts. Others went to the harbor to listen by radio to the rescue attempts of the "Emerald Isle." For people who had seen so many of their own go out on the lakes the loss of the "Bradley" had more than just momentary excitement. It carried deeply personal and bitter experience.

It is a human attribute to dwell much more on the rare and dramatic shipping disasters than on the uneventful and efficient carrying on of business on the wide waters of the lakes. For this reason we know much more about the lives and deaths of two of the many captains that Beaver Island has given to the Great Lakes Fleets. They were Peter Kilty and Emmet Gallagher who both suffered the misfortune of losing their boats and their lives in lake storms.

Peter Kilty's career might almost be taken as the prototype of that of a Great Lakes sailor of that era. He was born at Saint James in 1860 of an Irish family that had preceded the Mormons to the Island. The family moved to Onekama in 1880 where Kilty worked on fish tugs and schooners and then moved up to jobs on passenger boats, eventually commanding the "Petoskey" which ran between Chicago and the northern Lake. In the 1890's the cross lake railroad car ferries came into existence and Peter Kilty was given command of the first one to be built, "Ann Arbor No. 1." After a few years he joined the Pere Marquette fleet and moved his family to Ludington. By 1910 he was Pere Marquette's senior captain and was greatly respected as a sailor and man.

In September of 1910 he took "Pere Marquette No. 18" out on its regular 92 mile run from Ludington to Milwaukee with a load of 29 freight cars. He encountered a fairly strong north wind and heavy seas but not really anything too severe for his rugged type of vessel. Half way across, an engineer discovered seven feet of water in an aft compartment. It was disclosed that some of the deadlights had given way and that the pumps could not keep up with the inflow of water. A dozen freight cars were run off the stern in a desperate attempt to lighten her. "No. 18" and her sisters were some of the rare lake vessels of those times to be equipped with wireless and "No. 17" was called to assist. Captain Kilty thought that they might be able to make the port of Cheboygan but almost without warning "No. 18" plunged to the bottom taking 29 of her crew with her. Captain Kilty and all his officers were drowned. "No. 17" picked up 35 survivors.

Kilty was subjected to some criticism for the heavy loss of life because he failed to abandon ship. It was thought that he considered the danger of attempting to transfer people to small boats in the face of the heavy seas running was greater than the danger of his vessel foundering. He made a command decision of the kind that lake captains often had to make in the face of storm, fog, and ice and paid for his decision with the loss of his boat and his life.

As a result of the loss of "No. 18," changes were made in car ferry construction and there has only been one serious accident resulting in loss of life since. Many Islanders have served in the now declining Ann Arbor, Grand Trunk and Pere Marquette (now Chesapeake and Ohio) fleets and some still do.

Emmet Gallagher was the principal actor in a tragedy that took the lives of more Beaver Islanders than the "Marold" explosion. Emmet belonged to one of the numerous Island Gallagher families that gave many lake captains to the fleets. The year 1924 found him commanding the bulk carrier "Clifton." He had as his second mate Chine McDonough and as two of his wheelsmen Peter Burns and Joe Scheid, all Beaver Islanders.

The "Clifton" was one of the Alexander MacDougall whalebacks and was heir to all the discomforts of that odd type of craft. It was also a very early self-unloader. About half the crew of thirty were Sturgeon Bay men. The "Clifton" loaded stone at Sturgeon Bay and on September 18, 1924 set sail for Detroit. After passing into Lake Huron Captain Gallagher found a freshening southwest wind which increased as the day wore on. An upbound vessel noticed that the "Clifton" was taking a beating while passing Forty Mile Point. That was the last that anyone saw of her. She went unreported at Port Huron and, sometime later, bits of floating wreckage were found and were identified as hers. It was thought that either the "Clifton's" hatches were breached or that her new self-unloading gear had rolled her over. Beaver Island and Sturgeon Bay went into mourning.

Another Island Gallagher was lost in the greatest lake storm of this century. This was John Gallagher who, in 1913 was an engineer on the large, new ore carrier "Henry B. Smith." This was one of nineteen boats destroyed in the monster storm of November 7th to 10th 1913, with the loss of 248 sailors.

John was one of the Gallaghers who was born on Aran Mor in Ireland and came to Beaver Island as a young man. While on the Island he was known as Johnnie White. He married Maggie Gallagher, daughter of big Neil and, near the end of the century, the couple moved to Escanaba out of which port John practiced the profession that was to lead him to his death thirteen years later.

There are good reasons to believe that Johnnie Gallagher and his 25 shipmates lost their lives as a result of shipping company politics, for the "Smith" instead of going down fighting to reach shelter, actually left Marquette harbor at the height of the storm while the crew were still frantically trying to secure the hatches. It seems that Captain Jimmy Owens had had a hard luck year on the "Smith." Minor accidents, loading delays and just plain bad luck resulted in his tonnage quota being way down. He was apparently so enraged by front office criticism that he took his boat out of Marquette harbor as soon as it was loaded with ore and headed for the Soo, showing no regard for such details as 80 mile an hour winds and 30 foot waves. Soon he changed his mind and set course for the shelter of the Keweenau Peninsula. The "Smith" disappeared in a welter of wind-blown snow and was never seen again. John Gallagher's body was the only one from the "Smith" ever recovered, six months later and a hundred miles away on Michipicoten Island. Details concerning Johnnie's life and death were communicated to us by his granddaughter, Mrs. W.R. Lehmann of Ishpeming.

Beaver Island has given a large number of sailors who reached the top of their profession in the lake fleets. The Captains: Some of those of more recent years have been Clifford White, Forrest Pratt, Edwin McCann, Andrew Green, Jack Burke, Conn, Owen and Clyde Gallagher, Charles LaFreniere, Pete Green, Jack Bradley, Bud Left and Ray Cull. It is interesting to note that, at one and the same time, Conn and Clyde Gallagher commanded respectively, the fastest freighter on the lakes, the "Cliffs Victory" and the largest, the "E.T. Weir."

Times change; larger vessels, winter navigation, and more job opportunities on the Island have caused a decline in the number of professional seamen. However, dark clouds, high winds and rolling seas still cause Islanders to turn anxiously to the lake and memories of other storms may yet bring a tear to the eye. Old habits die hard.

The "Pere Marquette."

Bibliography

Strang, James. *The Northern Islander.*

Boyer, Dwight. *Ghost Ships of the Great Lakes.* Dodd Mead and Co., 1968.

Boyer, Dwight. *True Tales of the Great Lakes.* Dodd Mead and Co., 1971.

Hilton, George W. *The Great Lakes Car Ferries.* Howell-North, 1962.

Ratigan, William. *Great Lakes Shipwrecks and Survivals.* William B. Eerdmans and Co., 1969.

Heden, Karl E. *Directory of Shipwrecks of the Great Lakes.* Bruce Humphries, 1966.

Mills, James C. *Our Inland Seas.* A.C. McClurg, 1910.

Dancey, Thomas B. Article "The Sinking of P.M. 18" *The Telescope* Vol. 18 No. 5.

Letter from Mrs. W.R. Lehmann.

We appreciate being able to have drawn on the memories of Lawrence and Bud McDonough, Russell Green and Mary Minor.

James H. Dormer

THE PRE-MORMON SETTLEMENT ON BEAVER ISLAND, 1837-1852

by Helen Collar

The first white man to reach Beaver Island, except for possible French fur traders, was the missionary priest Father Baraga. Between his initial visit in 1832 and the beginning of the Irish fishing community in 1856 and 1857, the Island was to see several changes. Father Baraga reported to his sponsors, the Leopoldine Foundation of Austria, that the inhabitants of Big Beaver were all "pagan Indians living in eight miserable huts." This was soon to change for the Island was to become, first a trading center clustered around two villages of white traders and fishermen, and later the site of a theocratic experiment by dissident Mormons who came to Beaver intent upon creating their version of man's ever recurring dream of Utopia.

No such lofty sentiments as creating God's Kingdom on earth motivated the first white settlers since, like most American pioneers, their ambitions were practical and secular. Their hopes for the future centered around money; they sought new homes in order to better their economic position and thus to make a fuller life for themselves and for their children.

It is difficult to find accurate information about the number of pre-Mormon white settlers on Beaver Island. Mr. M.M. Quaife in his *Kingdom of Saint James* wrote that when the first Mormon party arrived in the spring of 1847 the white population of Beaver was limited to the men at the trading post of the Northwest Company of Rochester New York, which, he said, was located on the north side of the harbor. This is incorrect. The store and warehouse of this company, under the local management of Erri James Moore, who called himself "warehouseman" in the 1850 census, and Randolph Dinesmore, who gave his occupation as "merchant," was located, not on the north, but on the west side of the harbor. On the north side were the dock, store, cooper shop, and home of Alva Cable from Fairport Ohio. This establishment, very soon after the coming of the Mormons, passed into the hands of Peter McKinley, also of Fairport. At the same time on the north side of what today we call Cable's Bay, was located another trading establishment owned and operated by Alva Cable's nephew, James Cable.

There was possibly a fourth trading venture situated near the Indian village on Indian Point. I have been told by an old timer, both of whose grandfathers reached Beaver in 1857, that "the first store on the Island was at Pagetown." The area called Pagetown, so called because of Mormon families named Page, is located near the first property on Beaver to have been bought from the United States government. In the summer of 1848, when Island land was first put on the market, a David McKay filed for this tract under the Pre-Emption Act. Putting the fact of McKay's early ownership, based on pre-emption rights, together with the Island tradition that the location of the first store was near the Indian village, I think it reasonable to speculate that David McKay was an Indian trader who may have antedated the Cables. We have no information about McKay except this land record, for the 1850 census taker either missed him or McKay had left the Island by that year.

Strang usually referred to the Islands as uninhabited, the place selected by God as a homeland for His Chosen People, those Saints who acknowledged the leadership of His ordained Prophet, James J. Strang. In his *Ancient and Modern Michilimackinac,* published in 1854, Strang did admit that there had been a few people who preceded his Mormons to this promised land; but he dismissed them as follows —

> "At this time (1847) there were three white families temporarily settled in the township, four or five men with Indian wives, several French half breeds, and twenty or thirty single men whose families were elsewhere, spending the winter at Beaver."

The earliest information I have been able to obtain has come from the descendants of James Cable of Cable's Bay. In the spring of 1837 Alva Cable and his 16 year old nephew James came to Michigan and, with a small vessel, began going up and down the east shore of Lake Michigan from Saugatuck to Pine River (Charlevoix) fishing and trading. During the summer they stopped at Beaver Island and there they found a few white men living on the north side of the Harbor on the point where the Harbor Light now stands. Thus we know that before James Strang first set foot on the Beavers in 1847 there had been a small settlement existing at the Harbor. A more stable community was soon to be established, for Alva Cable decided that Beaver Island was a good place to locate a permanent business, and the next spring (1838) he and James returned, building a store, dock, and home.

We do not know with certainty the identity of the men the Cables found in 1837, or just when these men had come to the Island. When, in 1832, Father Baraga sent his first report about Beaver to the Leopoldine Foundation, he made no mention of the presence of white men. Therefore, it would seem that the men found by the Cables must have come sometime during the five years between 1832 and 1837.

We can speculate as to the identity of at least two of these men. One could have been the David McKay whom we have already discussed; another might have been Eber Ward. There is a curious entry in the land records in the Charlevoix County Court House. On June 26, 1837 the United States Government paid Eber Ward $50.00 for his claim to part of the NW¼ or Lot 2 of Sec. 26 T39N, R10W. This was only 15 months after a treaty had been signed between the United States and the Chippewa and Ottawa tribes ceding Indian rights to land which included Beaver Island. There is no record in the Tract Books of the Bureau of Land Management, Department of the Interior, of this transaction between the United States and Eber Ward but some sort of an exchange of money must have taken place or it would not be in the County Records. Moreover when James Dormer, on June 21, 1869, bought Lot 2 he paid Eber Ward and his wife $200.00 for a Quit Claim Deed to the property. It is safe to say that when the Cables first reached Beaver Island in 1837, Eber Ward had preceded them.

When Alva and James started their business in 1838 they were following a trend that was developing in the region. The fur trade had passed its peak in the 1830's and fishing was becoming more and more the industry of northern Michigan. *The History of the Grand Traverse Region* described that growing business.

> "Some who came to the country in those early days to fish remained as permanent citizens; but generally the

> fisherman was a transient person, establishing himself anywhere on the shore where there was a promise of success in his pursuit, and readily changing his location as immediate interest seemed to dictate. Small trading establishments sprang up at various points, drawing their custom from both fishermen and Indians. A few small vessels, or 'hookers,' found a lucrative business trading from place to place, selling supplies and purchasing fish."

In this account the "transient fisherman" seems to describe the type of man found on Beaver by the Cables when they sailed into the harbor in their "hooker"; Alva's establishment the next year was typical of the more permanent trading centers of later development.

It is interesting to read Strang's opinion of these fishermen and traders written in 1854.

> "The fishermen, until within a few years, were all Indians and Frenchmen continually in debt to the traders. Gradually a few Americans and Irish went to the fisheries. Some of these took with them small stocks for trade. This taking intoxicating liquors among the Indians made their use more common and fatal. These men were bred to civilization and had gone among savages to get beyond the restraints of the law. They were the worst class of men, stealing the catch of the Indians at night and selling it as their own. Since 1843 merchants and traders have established themselves at other stations than Mackinac, more convenient to the fisheries. Such were the habits of dissipation on the fishing grounds the fishermen were worse off at the end of the season than at its beginning and were destitute of credit. This threw them into the hands of felons and outlaws."

Although Strang must be considered a prejudiced observer of his rivals for possession of Beaver Island, the fact that where The Harbor Light now stands was called Whiskey Point by the earliest settlers may have some bearing on the accuracy of his appraisal of his predecessors. The Prophet went on to describe —

> "A new class of fishermen (by which he seems to mean men of his Mormon Community) are persons of limited means, temperate habits, good morals, and persevering industry, from the best sections of the northern States and Canada who have come to the country to make a permanent home. They farm or establish mechanic shops

when not fishing, and conduct their business as in the best regulated, civilized societies."

The Cables were not the only traders at the Harbor to develop a permanent business. In his *Island Life* Robert Hatt wrote, "The American Fur Company had a building at the harbor about 1830." In no other source have I read or heard that this company, owned by John Jacob Astor, had a station on Beaver Island. It would seem unlikely that Astor's Mackinac Factors would set up a new post in the waning days of the company's existence, a company which was dissolved in 1834. Father Baraga made no mention of such a building at the Harbor, and I can not help thinking Mr. Hatt confused Astor's company with the Northwest Trading Company of Rochester, the date of whose first establishment on Beaver is uncertain. One whose grandparents were among the first to reach Beaver after the Mormon Exodus has told me, "The Northwest Trading Company built their trading post before 1841." That will have to be as close to a date as we can get; neither Erri James Moore nor Randoph Dinesmore, resident managers, are to be found in the 1840 Census rolls; but the Company was well established by the time Strang corresponded, in February of 1847, with its owner, Colonel John Fisk.

One of the most lucrative parts of the business carried on by all traders on Beaver Island, with the probable exception of McKay whose property was inland, was the selling of fuel to the growing number of steamships plying the Lakes. The side-wheeler "Walk-in-the-Water" had been the first steam powered boat to sail above Niagara. Launched on Lake Erie in the summer of 1818, it was 1821 before she entered Lake Michigan. With the opening of the Erie Canal in 1825, trade on the Lakes increased, but the real start of steam navigation on Lake Michigan did not come until 1832. By 1836 Gurdon Hubbard had established the first regular shipping service between Chicago and Buffalo, the Eagle Steamship Company. The boat "Erie" on which the English traveler, J.S. Buckingham went from Chicago to Buffalo in the summer of 1840 was probably one of Hubbard's ships. Mr. Buckingham has left us a description of a refueling stop. The "Erie" docked at South Manitou Island. What took place is undoubtedly the same operation which occurred on Beaver Island over and over again, both at the Harbor and the Cable's Bay

"At 3 p.m. we went to the anchorage of the South Manitou Island to procure a supply of wood for the boat. The passengers all landed to take a ramble through the woods. . . . The whole Island was occupied by a squatter, who was realizing large profits by the monopoly he enjoyed of supplying fuel to the steamers passing up and down the Lake. He employs about 20 men as wood cutters; and as the great distance from hence to Chicago on the one hand and to Mackinac on the other, 200 and 100 miles respectively, renders it indispensable that the wood should be taken in here, he has his own price and charges $3.00 a cord for the same fuel as can be had at either of the places named for $2.00, because its bulk will not allow of the boat taking sufficient to run the whole 300 miles without lumbering their decks as to lose the requisite room for passengers and cargo. The boats *must* have it here, therefore, and hence the high price demanded. The Erie, in which we were, a boat of about 500 tons, consumed a cord and a half per hour; and the wood she required for her whole trip from Chicago to Buffalo and back, at the varying prices at different points, about $500.00 for a distance of 1000 miles."

The monopoly enjoyed by the squatter of South Manitou was to end, for in October of 1848 Strang wrote in the Voree Herald that wood sold at St. James for 50 cents a cord, and that sixty lake steamers had called during that year's shipping season. The "Erie" went near the Beavers as she continued north toward Mackinac, but she did not stop at the harbor. Mr. Buckingham continued —

"At sunset we were opposite, or rather between the Great and Little Beavers. . . . There is still an Indian settlement on Big Beaver; the people living principally by fishing and occasional traffic in peltries."

The author must have received his information from sailors on the "Erie," and he heard nothing of the new trading post that had been built two years earlier by Alva Cable on Whiskey Point. Little Beaver is now called Garden Island, and this account verifies what was told me many years ago, that the main channel used by the boats going from Chicago to Mackinac had been, in early times, between Beaver and Garden Islands. As his ship passed between "Great and Little Beaver," Mr. Buckingham continued his diary, commenting upon "the unusually beautiful sunset, the lake like glass, and the heavens decked in the most glowing tints toward the west, and these reflected in the clearest mirror." When these words were written the "Erie" was passing in front of the beach on which my summer home now stands. Many things have changed in the 140 years since Mr. Buckingham saw his "unusually beautiful sunset," but "glowing tints toward the west" are still reflected "in the clearest mirror" as one looks on a summer evening across the twenty miles of water that lie between Beaver Island and Michigan's northern peninsula.

There are no records to show the population of Beaver, either Mormon or Gentile, before 1850. We can only conjecture as to the number of people living there when Strang and his four companions came to assess the Island's possibilities as a future home for the Saints. The first indisputable correct information about individuals living there is contained in the Land Office Tract Books of 1848 when Island land was put on the market. Seven Gentiles brought acreage that summer: David McKay, James Cable, Alva Cable, Erri James Moore, Randolph Dinesmore, Thomas Bennett and Patrick Looney. The only Mormons buying from the government that year were Marvin Aldrich and Daniel Adams, but Mormons Bowers, Dodge, Cole, and Mills bought from Moore small tracts of the acreage he had purchased.

Another source of information about individuals is Elizabeth Whitney Williams' book, *Child of the Sea,* written in her old age. Mrs. Williams came to Beaver Island in the spring of 1848 at the age of four or five. Not only her youth at the time about which she wrote, but also her extreme prejudice against the Mormons, make her recollections of doubtful value. However, I think that when she wrote about her family's neighbors, we can accept as fact their presence on the Island. The Whitneys first lived in the harbor area near the elderly couple Mr. and Mrs. Patrick Looney and the Looney's nephew Patrick Kilty. Soon after the Whitneys' arrival the Alva Cable property changed hands. Alva went to Cable's Bay to join his nephew and Peter McKinley took over the business on Whiskey Point thus becoming, with his wife, his three daughters, and his sister Ann, part of the Gentile community at the Harbor. Besides Thomas Bennett, his wife and three children, the Whitneys had as a neighbor Tom's brother Samuel. Both Walter Whitney and Thomas Bennett fell into Strang's category of "a few men with Indian wives." Also living at the Harbor were Erri James Moore with his wife and Randolph Dinesmore, both men agents of the Northwest Trading Company.

At Cable's Bay lived James Cable and his wife and, by the fall of '48, his uncle and aunt, the Alva Cables. Mrs. Williams also wrote of the cooper at Cable's Bay, C.R. Wright.

Although these are the only non-Mormon names we can cite with certainty as being on the Island in 1848, there were others. There must have been more coopers than the Mr. Wright of whom Mrs. Williams wrote if the needs of three trading posts were to be cared for. These posts must also have had clerks, wood cutters to provide the cord wood for passing steamers, and the fishermen needed as a basis for their commercial operations. In fact, the reason Mrs. Williams gave for the Whitneys' move to the Island was —

> "Hearing that there were many people on settling Beaver Island, several families that we knew from New York State, Ohio, and Canada, my father made up his mind to go there."

One family, known to the Whitneys, whom we have reason to believe was on Beaver in early Mormon times was that of Joe Palmer and his wife Anna Williams. Mrs. Elizabeth Williams, nee Whitney whose second husband was probably related to this family, wrote of the friendship between her mother and Anna's sister, Abigail Williams Courchane, when both were living on St. Helena's Island just before the Whitneys' move to Beaver. Mary Palmer Floyd, who lived most of her adult life in St. James, was the daughter of Joe Palmer and Anna Williams, and she remembered seeing King Strang on the Island when she lived there as a child. Thus her childhood memory places the family on Beaver sometime before the exodus of the Gentiles in 1852. The conjecture of the Palmer's early Island residence is strengthened by our knowledge that the Palmer family and Mrs. Palmer's brothers and sisters were among those who, like the Cables, the McKinleys, the Looneys, and the Kiltys, returned to Beaver immediately after the expulsion of the Mormons. This is the only family known to have been Islanders in Mormon times, some of whose descendants still make their homes on Beaver.

Turning to the Mormons, we find that they were few in number when the Whitneys landed in the spring of 1848. Only four families of Saints had spent the preceding winter on the Island, but by the next winter, that of '48-'49, the number had increased to twelve families. Summer brought re-inforcements, and the winter of '49-'50 found a community of fifty Mormon families. During that winter Strang launched an intensive missionary effort in the eastern States, which resulted in a large number of converts joining the Beaver Island community. In the spring the final move of his headquarters from Voree, Wisconsin was accomplished, and the formation of his kingdom formally established by his coronation on June eighth before about three hundred devout followers.

With the coming of the census taker in the October following the coronation, we have our first reasonably accurate population figures. The 1850 census shows 483 people living on the Island, 32 households of Gentiles containing 128 people; 60 households of Mormons with 355 members. The recent summer migration of Mormons had raised their number until it far exceeded that of the Gentiles. Beaver Island's population in October of 1850 was 74% Mormon.

By this time almost all Beaver's non-Mormons had left the harbor area, and the census taker found them concentrated at the southern end of the Island, most of their homes clustered around Cable's dock and store. The only Gentiles still living on the Harbor, besides the household of Peter McKinley at Whiskey Point, were Erri James Moore and his wife, Randolph Dinesmore, Richard Cooper, and Henry and Mrs. VanAllen. VanAllen was soon to move; when the tower and the house at the Head were completed, he became the first Keeper of the Beaver Island Light.

Mrs. Williams stated in her book that the Whitneys, the Looneys and the Bennetts had moved from the town of St. James because they feared their Mormon neighbors who were showing a growing arrogance as their numbers strengthened. This account of Mormon pressure against non-believers is borne out by traditions handed down in both the Cable and the McKinley families. It is only fair to say that when, earlier, the Gentiles had outnumbered the Saints, they too had exerted unfriendly pressure against those they considered intruders, whose ways they did not understand, and whom they did not want as neighbors. But in 1850 it was the Mormons who were on top of the heap, and most of the Gentiles felt it was safer to leave the Harbor to the newcomers, moving themselves and their families to Cable's Bay where they were among their own kind. Left in the Mormon town were only those original settlers with a heavy investment in buildings and real estate, McKinley, Moore, and Dinesmore.

The McKinley household on Whiskey Point deserves special mention because of its size. Besides Peter's wife Ellen, their three daughters, and his sister Ann there also lived in this house a cooper named Charles McGuire with his wife, three other coopers who were single men, two clerks who were also single, and one man whose occupation was listed as "none," seemingly a boarder. At Cable's Bay, James Cable had a similarly large establishment. These two businesses had been founded many years before the coming of the Strangites, and their size shows that the pre-Mormon population must have been much larger than the enumeration of known inhabitants whose names we have been able to list.

Turning to Cable's Bay, we find there in 1850, a village of 27 households with 108 people, all non-Mormons. The largest group living together was the James Cable menage. This household embraced, besides James, his wife and son, his father, his sister and her husband, his brother and six single men who included three coopers, one fisherman, one farmer and one shoemaker.

When we look at the occupations given in the census we see a marked difference in the character of the two settlements. By far the largest number of Mormon men gave their occupation as farmer, forty-one of the eighty-four polled. Only ten were fishermen and three were coopers. Eleven men worked with wood, including four lumbermen, five carpenters and joiners, one cabinet worker, and one ship's carpenter. There were three Mormon merchants and three ministers. Four men dealt in services, of whom three were shoemakers and one a tailor. The rest of the men called themselves laborers and seamen. The King gave his occupation as "lawyer."

The non-Mormon community was a fishing village. Of the sixty-eight men employed, forty-six were fishermen and ten were coopers, a trade that was a necessary corollary of fishing. Thus 82% of the men at Cable's Bay depended upon the fishing industry for their livelihood. Four men were merchants, two were clerks, two farmers, and one each were shoemaker, trader, ships carpenter, and warehouseman. Two listed their occupations as "none."

Many of the Gentiles were not considered by Strang as residents of the Island, for some, but not all, spent the winter months in Ohio or New York. If a man's home may be considered the place where he spends the greater part of the year, these people were indeed Islanders. According to all accounts, they came with the opening of navigation in the spring, usually in early April, and did not leave until winter closed down in November. They brought their families, for out of thirty-seven households of Gentiles on the Island, in twenty-six were wives or grown daughters as housekeepers. Although many of the fishermen were young, recently married, and so far childless, there were thirty-seven children in the community. In fact, the reason given by some families for their 4½ month winter sojourn elsewhere was to get adequate schooling for their children.

A non-Mormon community of this size and stability did not develop during the three years between the coming of the Mormon exploratory party in the spring of 1847 and the fall of 1850 when the census taker arrived. This seems particularly true when we consider that, in the latter part of those three years, non-Mormons were under constant pressure to leave their homes, as witnessed by the almost total move made from the Harbor to Cable's Bay. Strang's uninhabited Island must have been harboring a larger population than the King conceded with his "three white families, a few half breeds, and twenty or thirty single men with families elsewhere"; or than Mr. Quaife postulated with his "white population limited to the post of the Northwest Trading Company."

What happened to this settlement of non-Mormons, made up, according to the 1850 Census, of thirty-two homes and one hundred and twenty-eight people? During the next two years, from the taking of the census to the close of navigation in the fall of 1852, all non-Mormons left Beaver Island, leaving behind homes and business investments. Friction between the two communities had mounted and pressure had increased against the Gentiles, until their position had become untenable. The events that led to this can be briefly summarized.

Before 1850 any persecution that had been practiced on the Island had been perpetrated by the original settlers against the unwanted Mormon newcomers. Strang, in his book on Mackinac, wrote about the low character of his neighbors, calling them thieves and felons. He told of Mormons being beaten by fishermen when they went to Whiskey Point, of efforts to break up Mormon meetings by "rude and filthy" conduct, of the stealing of a cow given by Strang to a "poor widow woman." As he cannot be classed as an unprejudiced recorder of the friction between Mormon and Gentile, Strang's writings are to be viewed with some scepticism. It must be remembered, however, that this was a frontier community and the ways of frontiers are rude and direct, with little regard for the niceties of life or for the law. The Mormons came into a typical fishing and trading settlement with its own customs, and Mormon ways were not their ways, and therefore suspect. As the number of Saints increased, with no end in sight, dislike turned into fear and fear bred action. Probably there was a basis of truth in the charges Strang made about his neighbors, and acts of violence were perpetrated against the Mormons, but at this distance it is impossible to distinguish in the morass of Strang's often intemperate language between truth and exaggeration.

One authenticated fracas between the adversaries took place on July 4, 1850. The heavy migration of Mormons in spring and summer had alarmed the Island's original residents. Calling on fishermen from Mackinac and other nearby settlements, the Gentiles gathered at McKinley's store on Whiskey Point on July 3rd, planning to invade the Mormon town across the bay and celebrate the nation's birthday by an attack on those they considered theirs, and the nation's, enemies. Perhaps they remembered that in March Strang's second in command, G.J. Adams, had prayed, "Oh God! Hasten that time that such a cruel, blood-thirsty nation may be blotted out from under Heaven, and be no more a people."

Strang knew of the plans of his enemies and had made preparations. When dawn broke on the fourth, the fishermen assembled on Whiskey Point, who had celebrated their expected victory the night before were startled out of their sleepy hang-overs by the sound of cannon fire. Across the Harbor the Mormons, emboldened by the strength of their increased numbers, had secretly imported and installed an artillery piece and were getting the range of McKinley's buildings. Soon a delegation of Saints arrived to notify McKinley that he would be held responsible for the actions of his visitors. A sobered and crestfallen group of men returned to Mackinac and to St. Helena's Island and the Battle of the Fourth was over. Later, on July 8th, Strang was proclaimed King, his crown placed on his head by the Apostle Adams, who in March had prayed for the destruction of the United States.

With the Kingdom established and the Mormons now outnumbering the Gentiles three to one, Strang publicly declared that the Saints would turn the other cheek no longer. This resolve was strengthened the next February by a revelation from God. Through his prophet James, the Almighty spoke,

> "Let your fear be upon all men . . . for this is the day of the vengeance of the Lord and of your recompense upon your enemies. Arise and thrash, for I will make thy power iron . . . thou shalt break to pieces many people, and shall consecrate their spoil unto God."

This revelation was the basis of the widespread belief among non-Mormons that Strang and his followers considered the appropriation of the property of others to be not robbery but "consecration." The Author of *The Grand Traverse Region* stated, in 1884, as a fact that —

> "The consecrating of Gentile property, or, in other words, the robbing of those who were not Mormons, was a recognized and established practice from the earliest settlement of the Island until Strang's death. It was the natural and legitimate sequence of the doctrine that the Mormons were God's peculiar people, who alone had the right to the earth, and were eventually to possess it."

Certainly, by all accounts, for the next few years a great deal of thievery took place on and around Beaver Island. Fish were stolen; boats and nets mysteriously disappeared, and the general belief throughout northern Michigan was that this was the work of the Mormons, spearheaded by an organized band called "The Forty Thieves" sailing a boat named "the Pirate Ship." I even own a piece of wood given my father said to have been taken from the sunken hull of the "Pirate Ship." This is all too romantic to have the ring of truth or to be taken at face value. Nevertheless we must believe that under the Gothic flourishes are some kernels of reality, if for no other reason than that Strang did protest too much. In *Ancient and Modern Michilimackinac* he wrote —

> "Many of the outlaws who had been engaged in the unsuccessful attempt to expel the Mormons (on July 4th) resorted to Grand Traverse Bay. . . . A gang of villains . . . ostensibly engaged in fishing, made frequent trips to the Head of Beaver, where, loading their boats with such plunder as they could sieze upon, they went with the same wind to some uninhabited island, and waited for a fair wind to return. . . . Those incursions were made at night . . . and their depredations were committed in the neighborhood of the Mormon setlements, and, because of the usual prejudices against the Mormons, . . . they gen-

erally succeeded in charging their crimes to them."

What seems to be another version of this same incident, the stealing of goods at the Head, occurs in the *History of the Grand Traverse Bay Area.* The author interviewed a man whom he does not name, but from his description must have been Austin Campbell, a Mormon who became an apostate, and who was allowed to stay on after Strang's assassination because of his known anti-Strang sentiments. Part of the interview follows, "the narrative given nearly in his own words." It is about "a man named Martin" who must have been one of the family of brothers from New York, Edwin, William, and Samuel Martin found by the 1850 census taker at the south end of Beaver. This family is not to be confused with the Martin brothers living at that time on Mackinac Island who moved to Beaver after the Mormon Exodus.

> "A man named Martin was fishing at the head of the Island. At the beginning of winter he packed up his property, hauled his boats on the beach, and left all in the care of one of his men. One Saturday Strang preached a very able sermon, advocating the right and duty of 'consecration'; and declaring he would not ask others to do what he would not do himself. The next day Silas Miller, who was a Gentile, and myself went to Sand Bay after hay. There we found one of Martin's boats, loaded with his own property stuck in the ice near the shore, and Strang and old General Miller drying themselves by a fire in a shanty. They appeared to have been in the water. . . . On returning to the village, Strang and General Miller sent out teams to bring in the goods."

One may believe whichever of these narratives one chooses, but evidence is strong that robberies did take place. Undoubtedly the law was broken by both Mormons and their critics, each finding the other a convenient scapegoat. It is in the nature of all frontier communities, both Mormon and non-Mormon, to have among their number some who are rascals and thieves, but in matters of robbery, the Mormons are particularly open to suspicion and vulnerable to blame because of Strang's announced policy of "consecrating spoils unto God."

In 1851 the Mormons used their newly acquired majority to effect the political organization of the Island. In 1847 Beaver had been designated as Peaine Township. Several meetings had been held but it had been found impossible to organize a government, probably because of the rivalry between Mormon and non-Mormon, neither of whom could call out a clear majority. In 1851 organization had taken place and all officers elected were Strangites. This meant that any local arrests would be made by a Mormon Deputy Sheriff and that any legal disputes would come before a Mormon Justice of the Peace.

In the spring of 1851 the Constable tried to arrest Erri James Moore for selling whiskey to the Indians. Moore, who had been clearly breaking the law, resisted arrest; shots were fired but no one was injured. The culprit fled to the more friendly jurisdiction of Mackinac Island and persuaded Justice Charles O'Malley to issue warrants against the thirty-eight Mormons who, he said, had tried to arrest him illegally. In the end all charges against the Mormons were dropped.

A few weeks after the Moore affair, the Mormon Samuel Graham, when he went to Whiskey Point on business, was set upon by the fishermen Richard O'Donnell and James Hay and severely beaten. An attempt by the Mormon Constable to arrest these two men met with resistance. A confrontation took place between the two factions at Cables' Bay ending in gun fire, both sides claiming the other fired first, When the smoke had cleared, the fisherman Tom Bennett was dead and his brother Sam wounded. Several Mormons were arrested for murder and taken to Mackinac Island, but in the end all were freed.

The most serious of these brushes with the law suffered by the Mormons occurred in the summer of 1851 in a Federal court in Detroit. This is a complicated story which cannot be treated adequately here. While visiting his brother in Detroit, President Fillmore ordered an investigation of charges that the Beaver Island Mormons were guilty of treason, mail robberty, counterfeiting, and trespassing upon government lands. One of the principal witnesses for the prosecution was Peter McKinley. Ably defended by their Lawyer-King, the Mormons were found not guilty and returned in triumph to Beaver.

Before winter closed in, McKinley took his family back to Ohio in order, so recorded his daughter in later years, that his three girls might attend a good school. With the opening of navigation in the spring of 1852 the family started back to their home. When they reached Mackinac Island they learned that their property had been destroyed by the Mormons, probably in retaliation for McKinley's anti-Mormon testimony in the trial of the previous summer. This was not only a severe monetary loss for McKinley; the news made him fear returning home. Renting a fishing outfit, he stayed on Mackinac Island, fishing during the summer and teaching school through the winter. The McKinleys did not return to Beaver Island until after the Mormon Exodus of 1856, when, according to an account left by daughter Sarah, "only two peach trees were left where our home had once stood."

Peter McKinley had been Beaver Island's Postmaster, and when Dennis Chidester was appointed to replace him on February 24, 1852, Mormon control of all government offices in the Township was complete. In the Fall election, by his manipulation of the monolithic Mormon vote, Strang had himself elected to the State Legislature. Thus one fourth of the geographic area of the State of Michigan was represented by a crowned King.

The Kingdom continued to grow as the migration of converts flooded the Island and pressure against all unbelievers became more severe. There was widespread belief in northern Michigan that this pressure was exerted by overt acts on the part of Strang. It has been reported that he levied a tithe against the Gentiles as well as against his own followers, and that, when his power became impossible to resist, he decreed that all on the Island must join his Church or leave. At this date there is no way to verify the truth of these assertions. We know only that, by whatever means it was accomplished, by the late fall of 1852 all non-Mormons had left Beaver Island.

Mormon apologists have contended that the Gentile exodus was a figment of the imagination of Mrs. Williams and her ilk; that there were non-Mormons living on Beaver Island throughout the span of Strang's kingdom. The following is a quotation from a letter printed in the *New York Daily Tribune* on July 2, 1853. It was written by Strang after he had been shown an article that appeared in the *Tribune* on June fourth, an article derogatory of Mormon activities on Beaver Island. In discussing his relationship with the Indians Strang wrote —

> "Last fall (the fall of 1852) the Indians left the other fisheries, where whiskey was sold, and came to Beaver, where it was not. . . . No whiskey being there, and no persons but Mormons and Indians, not a word was heard about stealing. The fact is significant."

There is another significant fact here besides the one to which Strang referred. It is that on July 2, 1853 Strang publicly stated that in the fall of 1852 none were living on Beaver Island but Indians and Mormons.

We have already heard the story told by the McKinley family about their forced relocation because of the destruction of their property. There are two other accounts of forced departure, those given by the Whitneys and the Cables. Mrs. Williams' story of the Whitney departure is detailed, but allowance must always be made for her anti-Mormon prejudice. She wrote that Alva Cable, who had earlier moved to Pine River, came in a boat to take the cooper C.R. Wright to the mainland settlement. This left seven families besides the Whitneys at Cable's Bay, James Cable having already gone. The first of November a messenger came from Strang with a letter for each family saying that all must come to St. James to be baptized into the Church of Zion or leave the Island within ten days. Before twenty-four hours had passed all, except the Whitneys, had left, taking with them all their possessions that could be packed into their fishing boats. Mr. Whitney sent for a larger vessel to come for his family. On the tenth day the boat came; the family's goods were still being loaded when armed Mormons appeared on the beach. Abandoning the last of their household goods, the Whitneys set sail for Pine River.

The Cable story has been handed down by word of mouth to the descendants of James Cable. A Mormon friend warned him that he had better get away as soon as possible. James, his wife, and three year old son Claude got into their Mackinaw boat and hailed a passing ship, the "Montezuma." At first the Captain refused to take the refugees aboard because he lacked accommodations for a woman but he was finally convinced of the urgency of the situation when the Mormon boat "Black Maria" approached. The Captain relented; the little family boarded the ship, and the "Montezuma" sailed through a November storm to Chicago.

When the winter of 1852 came to northern Michigan it found all of Beaver Island in the hands of Strang and his Saints. We can be sure that the departure of the Gentiles was not wholly voluntary; business men like Peter McKinley and James Cable did not willingly abandon what for the times must have been large investments. This exodus, ignored by Mormon apologists, was smaller in numbers and less cruel in execution than the later Exodus of the Mormons themselves, but it was an equally forced departure of an unwanted minority. By the middle of November the pre-Mormon settlement on Beaver Island had come to an end, and the King was alone with his subjects: the Prophet with his Saints. Strang was free to work out unhindered his version of the Peaceable Kingdom.

James Cable

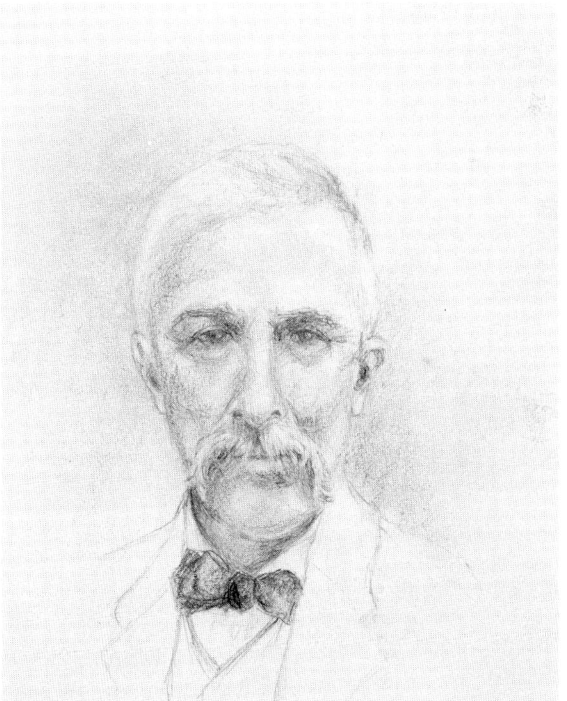

Claude Cable

Garden Island has a long and exotic harbor. The island is a short distance from the northern end of Beaver Island. In stormy weather the water is relatively quiet. There are rocky reefs near the harbor.

MINISS KITIGAN

by Keewaydinoquay

For people of Indian ancestry, Miniss Kitigan or Garden Island is more than a part of the Beaver archipelago; it holds rich associations with a culture now almost vanished. No written records exist, so what follows is necessarily limited; it is based upon fragmentary information from people native to the island, oft repeated legends, as well as my childhood and recent experiences of living on the island during summer months. A major resource person was my paternal grandfather's second wife, Minissing Odahniquay (Daughter-of-the Island) who was able to confirm many of my childhood memories when I was a summer visitor to Garden Island.

My paternal grandfather was a priest of the native religion, the Midewin. His name was Mideogema. He believed only in the old ways and refused absolutely to recognize any authority except the Divine Authority to which he knew he had access through the tenets of his religion. He travelled exclusively throughout the territory of "KewaydinAukee" (The Land of Keewaydin), which he believed had been given to the Algonquin peoples by sacred right and to which they had been led from the eastern coastlands by the Divine Megis.

Mideogema had two wives. My mother was wont to imply that the previously mentioned Minissing Odahniquay became his wife after the death of his first wife, but that was not the case; he simply had two wives simultaneously — much after the style of the Old Testament patriarchs — and both of his official wives came from Garden Island. My father recalled his father saying that the best women in the world came from Garden Island. Mideogema's first official wife was my bloodline grandmother. (If she had not died, my father would probably never have known which of the women had given birth to him. All the children called Mideogema "Nosse" and they all called both women "Niniga" without making any distinction.)

There is a ceremony in Indian custom by which formal adoption is made; this is usually used in the case of a step-parent, the unborn child of a lost warrior, or orphaned children of the same clan. According to this ancient Anishinabeg custom Minissing Odahniquay adopted my father for her own son and ever afterwards treated him as such.[1] Most of the time when my father said "Niniga" (my mother) he meant Minissing Odahniquay, and most of the time when I said "Nokomis" (my grandmother) I meant Minissing Odahniquay.

The first wife of Mideogema was called variously Ceceilia Moray, Ceceilia Manidu, or Mideogemikwe. Her father was a Scot by the name of James Moray, and her mother was a native woman of Garden Island called KishaWish, Cecilia Manidu, or Ceceilia Moray. Thus my father's blood mother was half Garden Islander and half Scotch. My paternal great-grandfather, James Moray (or "Murray"), was a Scottish merchantman whose boat was wrecked on the Whiskey Island reef. Grandfather said his father went to sleep in the cold waves of the reef and woke up on Garden Island in the warm arms of KishaWish, and saw no reason to move!

There were good reasons to respect this Scotsman; at a time when many traders had "squaws" and some native women were glad to accept such status, James Moray made KishaWish his legal wife according to the traditions of both their peoples. He lived on Garden Island and local environs for about ten years, becoming a man of considerable standing among his wife's people due to his agricultural knowledge and his talent for profitable trade with the whites.

This period of his life was interrupted by the news that his father in Scotland was dying. James Moray started the long journey to his homeland, hoping to arrive before his father's seemingly inevitable death. Ironically, James succumbed to a sailors' plague and his father survived.

Father, as a boy, received many presents from his great-grandfather in Scotland; one of these was a complete outfit in the Moray tartan. What a sight that must have been — a little Indian kid chasing about the woodlands in plaid and kiltie!

The new and the old world blended in another respect — story-telling. Most Indian traditions are handed down orally, from generation to generation. Given James Moray's cir-

---

[1] Such an adoption, though recognized as binding by all Anishinabeg folk, did not eclipse our blood relationship, through my natural grandmother, to the Manidu family, and it was because of that relationship, plus the same clan connection, that we went first to the Manidu home in the event described on page 33.

cumstances, it was easy for him to fall in with the custom of narrating life experiences around the evening campfire. Village elders would relate not only their own experiences, but those of fathers, grandfathers and great-grandfathers. Similarly, James would tell of his homeland, describing his father's place in such detail that it was still being re-told in my father's time. If chance placed me on the North Sea coast of Scotland today, I could locate the Moray home.

Story-telling was not confined to the men. Minissing Odahniquay was a great story teller; she enriched my childhood summers with her entertaining tales of people living on Garden Island when she was a child, and of the many legends transmitted through the generations. Whenever we visited my grandfather's camp, which was frequently and for long periods of time, evenings were spent around the campfire singing, telling stories, and popping the seeds of the yellow pond lily. So clearly stamped on my childish mind were her words that, forty years later, I was able to locate such places as Sunning Rock, Death Hole, Keewaydin Point, Ever-Fishing Rock, The Net, and our family burial area with very little effort.

As already mentioned, my great-grandfather James Moray brought to Garden Island a considerable knowledge of agriculture. The island itself provided a fertile site; everything still grows luxuriantly. But it is the corn crop which was culturally, historically and anthropologically significant.

From time immemorial the Islanders cultivated huge patches of corn and squash in clearings made in the interior of the island. For this reason, passing whites often called Garden Island natives the "Cornshocks," to differentiate them from the Indians of the other islands. For this reason, the chief from the Beavers who visited Grand Traverse in 1870

Alec and Esther Cornstalk.

was called "Chief Cornshock" by the whites. Later when white officials demanded a second name from Garden Islanders, many assumed they wanted the name given them by earlier whites and replied that they were "Cornshockers."

On the mainland in Charlevoix County today there are a few people who still carry the surname of "Cornshock" or "Cornstalk." There are many more whose official records of descent show ancestors who bore one of the "corny" names. Regrettably, there are many more Indian families who carefully delineate that they are DEFINITELY NOT descended from any Cornshockers. In the later years of the Island's occupation by native peoples, "Cornshocker" became a local term of derogation, referring to those who clung doggedly to their island home and continued to live in ignorance of "civilized" benefits.

Today, in the central part of the island there are numerous flat places which may well be the original cornfield sites. Even with the accumulation of fifty years of secondary growth, the trails of these openings are discernible. The fields are located along a natural, or long-established, water flow. The majority of these areas are now flooded by the beaver Under the waters of the beaver ponds, roads, paths, buildings and platforms are clearly visible.

Several things point to an ancient occupation of Garden Island by Indians. In the legends of the Crane Clan, who were led from the crowded settlements along the Bwating (St. Mary's River) by their totem bird, the Crane, they crossed the straits of MichiMiknikMacknackinog ("The-Place-of-the-Cutting-Off-of-the-Great-Turtle-Trail") and came first to Miniss Kitigan. It is said they were hospitably received by the Corn people but did not stay, because the Crane had promised them a land unoccupied. The Corn people had already been there for many generations and were protected by a very powerful "Manitou."

A further attestation to the ancient occupation of Garden Island is the extent of the island's cemetery, which is reputed by such recognized authorities as Hinsdale and Densmore to be the largest native burial site in the central United States. Geologically speaking, it is interesting to note that experts agree upon the path of Algonkian migration, which followed the retreat line of the last glacial deposit of earth and stone from which the Beaver archipelago was formed. In recent years, islands in the same chain of geological formation have been used by agricultural specialists to develop new species of berries and grains. Red winter wheat was developed on North Manitou, Horstmann's dewberries on North Fox, and Brother Berrien's barley on High Island.

Childhood Memories

Much of my own early childhood was spent along Cathead Bay, which is now a part of Leelanau County. From the beach near our cabin of squared chestnut logs we looked directly at South Fox Island. On one occasion an Anishinabeg traveling to the lumber camps stopped at our place to tell my father to come immediately for a funeral feast on Garden Island. An earnest family debate followed. It was finally decided that we were under obligation to go as the woman for whom the feast was being held had once saved my father's life. At that time she had been the recognized Mashkikikwe of Garden Island. None of us could have guessed then that in the far years to come I would become the Mashkikikwe of Minnis Kitigan!

That visit showed me, for the first time, that my powerful father could know fear. We had travelled by canoe from our bay to South Fox without incident; on arrival my mother leapt from the canoe and danced up and down the beach of Miniss Shawaniwaug. That was a large expanse of open, dangerous water and her relief to be safely ashore was evident! The next stage of travel, from South Fox to North Fox was rough but passable. However, there was a mixture of fear and consternation on Wauboshtegwan's face, after landing on the marshy North Fox shore, as he strained to see the dim outlines of the Beaver group.

We passed over the turbulent strait and skirted the eastern shore of Garden, coming at last to the beautiful rustic landing of Peter Manidu, who was our clan cousin. My father said to my mother that Pete Manidu was a misplaced genius, recounting his many skills, which included horticulture, shipbuilding, and landscaping.

One of my mother's objections to the journey had been her desire to keep me from contact with the "pagan" religion, for she was the daughter of an Anglican missionary. Accordingly, I was left in the care of Peter Manidu's son, Joe Pete, who was not much older than my own seven years. He, too, had been told that he could not attend the feast, and neither of us liked the situation, but disobedience was not part of our upbringing. However, Joe Pete threatened to lock me into a small cellar under a trap door in the middle of the cabin floor if I didn't behave. Frightened, I ran off down the woodland cart trail in the direction my parents had disappeared. Joe Pete chased after me, but I eluded him by lying flat on my stomach in a large patch of bracken. He went back and forth several times and then returned to the cabin, leaving me to continue down the cart trail.

This trail led through deep woods to the lumbercamp triangle and from there to the Harbor of Refuge where the village of Miniss Kitigan was located. This is now called Indian Harbor, and the exact location of the old village is easily recognized.

At that time the North Cut had not been completed; on the trail, which seemed very long, I met people only once, a family group traveling in the opposite direction. I could see they had been to the feast, for they were carrying food away. The mother of the group came up to me making "mother" noises. She asked if I were lost, I answered that I was not lost and was going to join my parents at the funeral feast. She asked who my parents were and I told her. There was an old woman at the end of the group, very unkempt and dirty. Coming abreast of me she stopped and spoke to me; she had no teeth and a foul breath.

"I knew your grandmother, Ikweysens," she said. "I knew also *her* grandmother, your father's mother, and your father's mother's mother. They were good women, all of them, and I can see that you are truly their descendant. All of them had a great gift of healing. You must keep our heart happy so that you also can call forth the good spirits for healing." She might have told me much of my forebears, but at the time I wished only that the woman would pass along, even though my home training enabled me to face her courteously and directly. That chance meeting marked a turning point in my life.

Long before I reached my destination I could hear the sound of a great drum and the rise and fall of many voices. It echoed from the bay behind Little Island and sounded like two drums and thousands of people, a sound unknown which reached out to create a spirit throbbing in my blood. Rounding Catfish Cove and climbing the rise above the village, I saw a cortege of men bared to the waist carrying the body of my father's savior toward the northwest (the direction in which both the traditional and Christian cemeteries lie), and a multitude of people retiring the Drum to the East.

"Baum! Home!" the muffled, slow-measured Drum said. "Little-Walks-With-Bears-Woman, welcome to your heart's home. Bawm! Bawm! Home! Home!" This was my first sight of a large group of people moving rhythmically to a drum. I had never danced. My maternal grandfather felt dancing was sinful. But without hesitation I began dancing with a group of girls my own age who were in the procession retiring The Drum to the East. It was there that a greatly disturbed mother found me, scolded me thoroughly and laid on me the punishment of learning a chapter of scripture. It was duly learned — to the rhythm of The Drum, which still beat inside my brain. Nothing else mattered. It had Happened; The Drum has guided my life ever since.

* * *

In spite of their attempts to regulate my activity, Indian protocol kept my parents so occupied that I had much time to satisfy my naturally investigative nature. The Indian cemetery seemd an ideal place to explore. Here, the near relatives of the deceased were keeping the small spirit fire burning on the grave of my father's benefactor. The rest of the cemetery was available to me. I had never seen the likes of it before. Many of the spirit houses contained maple sugar cakes (good for nibbling!) on the little shelves just inside the spirit holes. Some had spirit flags between the wooden slats and these received a good wave! Children, the world over, seem to share similar impulses. Two plots had been enclosed by picket fences with wicket gates. Some of the spirit houses were surmounted by crosses, and some were constructed in "picket fence" style. Some were fancifully designed and all were painted bright white. They looked like an array of white doll houses. In this part of the cemetery were names such as Naganash, Kenoshaw, Wanigan, Johnson, Manidu, Nanigu, Pierre, LaCroix, Santee, Antoine, Baptiste, Shingwalk, Wawasee and many more.

Much of the non-Christian part was deep in forest gloom. Here sunken spots marked the actual burial sites. No names marked the burials, but the family totem was carved into the bark of living birch trees. Some families marked their plots with ingeniously bent trees; one birch was bent into a complete arch over the entire plot. Individual graves were marked with spirit houses made of logs, in the log cabin manner. In some places these spirit houses had almost disappeared into the ground. Beside many graves were intriguingly carved posts surmounted by animal figures in an upside down position. These were not of the pretentious proportions or detail of the North Pacific totems. The largest was no more than three feet in height, and the smallest about one foot. It was necessary to travel along a path of an upside down position in order to view the small animals! But for a child, this was fun.

One impressive spirit house was at least seven feet long and four and a half feet high. The apex of the roof towered over my head as I stood in front of it. But the person buried there had been very slight of build and small in stature. All his life he had been mockingly called "Little Joe Mahng." So he built a spirit house for himself that was larger than a chief's, and burned a sign for it which read "Big Joe Mahng."

In the part of the cemetery which borders the old road from the sugaring-off shed, the graves were all marked with stones. Indians of the old religion were always very careful not to put anything heavy above the deceased for they believed that this might hamper the movement of the spirit away from the body toward the Blessed Isle along the Spirit Trail Cycle. Accordingly, these stones were placed just beyond the head or the feet. Sometimes, particularly with babies or small children, the entire body would be outlined with stones. It is believed the people buried here were from the clan Ahmeek.

The Christian part of the cemetery was somewhat like other cemeteries, but not entirely. Most of it was in open clearings and illuminated by bright sunlight. On the majority of graves there were fresh flowers bunched in small containers and bottles buried in the ground. This gave the effect of hundreds of flower arrangements growing directly from the earth. It was amazing to see that the earth about most of these graves HAD BEEN SWEPT BY A BROOM! Those which hadn't been swept were made flat and level and smoothed by hand. Perhaps they were always like this or perhaps only at that time; however, the hundreds of them then made a great impression.

Some of the graves were marked by wooden crosses on which the name, age, and date of death were carved. Others were marked by fieldstones on which names had not too successfully been painted. Still others were marked by conventional and commercially lettered headstones. In reading all of them, I was struck by the large number of children's graves;

in my eyes at that time, dying was an act reserved for the elderly.

To the east of the main road which led from Indian Harbor to Larsen's Farm on the north shore, people of "no clan" were buried. Mainly the Indians had buried here bodies of persons unknown to them which had washed up along the shores of Garden and the nearby islands. These graves were marked with a single stake and frequently on or around the stake was some item found with the body; e.g. false teeth, wedding rings, religious medals, and store-bought shoes. A well-worn path to one item indicated its special fascination. It was a man's old-fashioned watch, very thick and heavy. Long since it had ceased to tell the time, but if you pulled a little button at the top, tiny chimes played a tune. The kids in the village told me that in the evening the spirit of the watch's owner came back to play the melody.

The number of graves did not compare to the number of people living in the village. The village was a small unit, much like Onominee near my home on the Leelanau, but the cemetery went on forever. Many years later I read in the Burton Historical Library that there were over 3500 graves.

Garden Island Cemetery.

Legends

The Indian names for the islands identify specific characteristics. Thus, the long, low island which is now called Hog Island by the white people was originally named *Miniss Ahsheegan,* "Island of the Bassfishes." Trout Island was called *Nahmay-Gosseekan;* "Place-Where-There-Are-Trout." Pismire Island was *EnigoWigamig:* "The Ant Family's Wigwam.[2]" High Island was *IshpesseneWangag:* "There-Are-High-Hills-of-Sand-Heaped-Up." Many stories relate to these islands and the special uses of them by the Indians. A favorite story concerns the "honeymooners."

The people of Garden Island called Gull Island Fishing Island. Except for the Village area and the cornfields, Garden was rather heavily wooded. This left very little space for drying fish, an activity obviously requiring open area and sun. Each summer a number of young couples and their families would go to Fishing Island to seine small fish and dry them on racks in the sun. Since the dried fish were divided among the community's families, the "honor" of who would work for the summer on Fishing Island was decided first by lot and then by turns. Honeymooning couples were the choice of persons to be sent for the summer to Ahmeekogenda to work the vegetable gardens. Here again it was simply easier to use land which was already prepared for the purpose than to clear more of the home island. On Beaver there were huge raised gardening beds with earthenwork walls around them. The plants raised there were mostly root crops which did not do so well in the stony shingle or wet muck of Garden Island, but the larger corn fields were always laid out in the interior of Garden.

"Did the honeymooners build the garden beds with the earthwork walls?" I asked of Minissing Odahniquay.

"Pfft, kaween," said Grandmother scornfully. "Little you know yet of young people in love. The Council was wise in those days. They knew that those young folk would have eyes for little but each other. It was the easiest job in the community."

"Well if the honeymooners didn't build the walled gardens, then who did?" I persisted.

---

[2] This could be translated variously: "The Ant House" or "AntHill' depending on whether your cultural background was English, American, French, or Indian. The present name is 17th century terminology.

Grandmother stopped her weaving with one rush poised in midair and a small frown of perplexity appeared between her wide brows. "I really don't know. No doubt it was the Good People back in the ancient days when The Hook was connected to Waugoshaunce. Now stop interrupting with your questions. I lose the rhythm of the story and miss out on the loops in my weaving. Your Grandfather likes an even mat under his feet." Further questioning by a child would be impolite, but I wondered, had Grandmother really heard from her people something about the mystery of who built the raised garden beds on Beaver Island? Nowadays anthropologists and archeologists would like to know the answer Grandmother did not or could not give.

A yarn with a more satisfactory ending concerns the "True-Hearted Woman" — a legend of Squaw Island. It was the custom, in the days when all of our people followed the native religion, for a woman to go apart from her family and the village at the time of her monthly flowing. When the weather was at all favorable, most of the local Anishinabikweg spent their "time apart" on the small island which lies northeast of Whiskey and northwest of Garden. For this reason it was originally called "Women's Island." (Not "Squaw" for that was a "bad word," meaning a woman who sold her body — or one who was sold or traded by her family or owner. It was European adventurers, trappers and traders who twisted the name.)

Over the years, an interesting little encampment was built up on Women's Island and one canoe was kept at Bomwe Bay for the use of the women. Contrary to what one might expect, this time apart in the women's encampment was greatly anticipated and enjoyed by the ladies. Even at its best, life was difficult for Indian women. Theirs was the drudgery which necessity, an austere climate, and a masculine-dominated society imposed upon them. This time apart afforded them relief from responsibilities and offered opportunity to rest and refresh themselves.

According to a tale which used to be told about the campfires of Miniss Kitigan, the name of the island was eventually changed to honor a special person who faithfully kept the tribal rules. Then it was called "Isle of the True-hearted Woman." It happened in this manner: True Heart was the young wife of an ambitious man, ambitious in the sense that he desired prestige, wealth, and easy living. He was NOT ambitious at working or hunting. This fact was noted by almost everyone but True Heart, who admired her handsome husband and did not seem to mind slaving to keep him in the style to which he wished to become accustomed. Her efforts were not entirely satisfactory to Ambitious Man however, for the extra efforts required of her soon made her thin and tired, and besides, she appeared to be barren.

On a trip to the mainland, the roving eye of Ambitious Man fell upon the attractive daughter of an Odawa drum chief. She was plump and pretty, and the social prominence of her father assured that a handsome dowry would become the property of her husband. Pretty Partridge seemed to enjoy the attentions of Ambitious Man, so he applied to the council of elders for permission to take a second wife. The council of elders were aware that the display of elegance and plenitude in the lodge of Ambitious Man was due to the extreme efforts of True Heart. (Their women had seen her sneaking off to the trap lines in the gray light of dawn so that no one would know Ambitious Man had not filled the cooking pots of his own household.) The council told Ambitious Man that he was not really supporting one wife, therefore, they could hardly grant him permission to have two women. Ambitious Man was enraged and frustrated. He was determined to have Pretty Partridge and her dowry.

Now it happened at this time that True Heart had gone over to the Women's Island alone for all of the other young women were pregnant. Only the medicine woman knew that True Heart also carried the hope of a child next to her heart.

"You must put on a sweet face and go to the Women's Island anyway," the medicine woman had told her, "for complete rest is the only way you can continue to carry this child you have so long desired. Keep very warm, put up your feet, sleep all you can. I myself will give you provisions for a nourishing broth laced with virtuous herbs."

When he supposed her time was up, Ambitious Man, feigning kindliness and consideration, took his own canoe to collect his wife from Women's Island. Fuming to himself in his lonely lodge, Ambitious Man had thought upon a plan which he now put into effect. Pulling up beside the women's canoe at the tying-up log, he slashed the little bark until it sank, then turned and paddled back toward Garden, leaving the unfortunate True Heart marooned.

All the villagers knew about the affair was that True

Heart's man had come back alone, bloody and bruised, daubed with the charcoal of mourning, and telling a strange story. The Women's Island, he said, was now inhabited by strange, vicious Matchi-Manidog (Evil Spirits) who wanted the island all for themselves. On arrival there he had found the body of his poor wife impaled and horribly mutilated. The Matchi-Manidog had released him from a similar fate only because they wished him to tell the Indians to stay away. Quite terrified, they did!

After the briefest possible mourning period for his supposedly dead wife, Ambitious Man brought Pretty Partridge and her dowry to his Garden Island lodge. The villagers were impressed with the dowry, but Pretty Partridge was not impressed with the island, the villagers, nor her new home. Without True Heart to keep it clean and attractive, the lodge had become filthy and dilapidated. Even Ambitious Man's beautifully beaded clothes and moccasins, with which he had cut quite a dashing figure on the mainland, needed mending. The lazy hunter had been eating up the winter's stores. There were skills which Pretty Partridge did not have, nor did she intend to demean herself by acquiring them.

It was not long before the whole village knew that Pretty Partridge was making a fool of Ambitious Man and that she could beat him at his own game of getting someone else to do the work. As the months passed, Ambitious Man lost his braggadocio and became shame-faced. The dowry canoes and the loss of prestige seemed a high price to pay for the presence of Pretty Partridge who flew his particular nest most of the time anyway. At last Pretty Partridge disappeared completely in the company of some fishermen from the Lost Lake Village on Pyramid Point, and the shame and degradation of Ambitious Man was complete.

He sat in the ruin of his lodge and ambitious plans and sulked. His emotions ranged between hate for Pretty Partridge who had made a fool of him and hate and fear of the spirit of his dead wife which he imagined to have brought this state of affairs upon him. Not once did it occur to this man that a little hard work on his part might alleviate the situation! There was only one solution, he reasoned, and this was to so placate the spirit of True Heart that she would not continue to visit bad luck upon him. Accordingly, he rummaged among the few remaining belongings of True Heart and found a daintily beaded bag which Pretty Partridge had missed. Of all that remained, this alone was in perfect enough condition to make a spirit offering. Her own handwork as an offering to her spirit! The incongruity of it was not apparent to the distraught man. As was the custom then, he marked his canoe in black and censed it with cedar, blackened and disheveled his person (which in his state was not TOO much work!), and set out, a sad sight indeed, intoning petitions to the zhee-be-ug (departed spirits).

On the island where he had once marooned his wife, he set about to peel a cedar tree as it was traditional to place manitou offerings on tall poles. A movement in the underbrush of a nearby swale caught his eye. Immediately the nervous man shot an arrow into the quivering bush. There was a piercing shriek! Ambitious Man dropped his offering and fled, but he did not go far. Heavy gutteral sounds pursued him, and he turned to face certain death by a manitou. It was no manitou. It was True Heart, and he knew, from the very human blood which gushed out where his arrow had pierced her abdomen, that she never had become a manidu.

"The child," she gasped piteously, "save your child . . ."

Ambitious Man remained bowed over the body of True Heart for many hours. He was stunned, fearful, remorseful, penitent, and then — mercifully inundated by grief.

The coldness of the rising night wind finally roused him to action.

"I shall build her a fine spirit house," he thought, "and then I shall die beside her. I deserve no more." A strange wail which was no coyote cut across his consciousness. In a trance, and still painted with his true wife's blood, Ambitious Man climbed a small rise and saw the rough shape of a bark wigwam. He peered inside. From a pile of rabbit skins, the frightened eyes of a wee boy child looked back at him. Again Ambitious Man was shaken with paroxysms of grief, and this time the quavering voice of the child joined in his wails.

It was a changed Ambitious Man who returned to Miniss Kitigan in the daylight. Dirty and disheveled, he seemed to have no thought for anything but tenderness for the strange child which he carried in his arms. Several women ran to his aid, but he would not surrender the child, nor would he speak until all had gathered in the center of the village around them.

Placing the child before them, he said, "This is my True

Son, the child of my true wife, True Heart." Then he told them all that had happened, sparing himself nothing. Great tears, like no proud Nishnabe would even allow to form, flowed unbidden down his face as he spoke, washing away the last blood of his true-hearted one. The old medicine woman stepped forward and of her knowledge verified his story.

"You may beat me to death now," said the grieving Ambitous Man, "but I beg you to raise the True Son of True Heart on Miniss Kitigan in the best Anishinawbimidisiwin."

Of course the people did not beat Ambitous Man to death. As quickly as they had condemned him before, now they supported him, and they did help him to raise True Son in the finest of the Indian traditions.

As for the brave True Heart, she was greatly extolled, and beside the fine spirit house that Ambitous Man had built for her, the people raised a big stone cairn. Each woman, every time she went to the island, laid a stone on that monument, with a prayer that she might be as courageous as True Heart. The men also remembered, leaving there, as they passed homeward, the stones with which they had weighted their fishing nets.

When the French traders came, they saw the native people leaving stones and said, "Why you do this?"

"It is an offering we make to the spirit of bravery."

"Then we will make one too," said the traders, "for anyone who paddles these waters surely needs bravery."

By and by the stones were so high that the great cairn could be seen far away across the water. Canoemen and sailors alike used it for a guide. Mothers, climbing the dunes of the nearby islands for kinnickinnick, pointed it out to their children. Fathers, showing their sons how to lay the whitefish nets, waved towards the great cairn.

"Get yourself a woman like True Heart," they would say.

"Enh," the sons would answer, "we shall seek a woman like True Heart."

When the steamers came, they used the stone cairn as a warning to steer clear of the currents which rushed to the reef on the other side. So for another reason, by another race and generation, the memory of True Heart was blessed. In later years, the government built a whole lighthouse of the many stones which had accumulated there.

Conclusion

There is an impelling mysterious spirit about Garden Island. Many have felt it, but no one knows exactly what it is — or if they know, it isn't being said.

"She calls her own, that one." Pete Manidu is supposed to have said this late one night in a Beaver Island bar. He ought to know. On Garden he buried the little FireFly girl, but he didn't leave. There he buried a much beloved wife, but he didn't leave. Over the grave where he expected to lie himself, he built the spirit house for his son, Joe Pete, but still he didn't go. When Pete's own time came, there was no room, so they had to bury him in the middle of the path under a cross he had made himself. He had not left Garden Island. "Like a woman she is, sends you away, and then calls you back."

Old Ben Shawboose, who made the first clearing and building at Golden's, finally left because of the Voices. Now his voice is one of those who call off Keewaydin Point. Anyone can hear them.

Grandmother Norton sleeps happily beneath the herbs she dispensed so wisely. Not only herbs were in her healing, but also spirit power. Nor did she go until she had called another to be mashkikikwe. Ask Keewaynah. Ask a thousand Irish descendants dispersed from the islands. She was midwife to their grandmothers. Or ask Nagoose, whose spirit lives again because of her.

Every fall Lettie Schlessinger walks the bog of Sorry Burn Lake with a bolt of autumn colored calico in her arms, and Golden Beard trims his apple trees, whistling answers to the birds and squirrels as he works.

And while the island waits, one family keeps the sacred pipe, and another family keeps the sacred drum, and all of us, in the spirit or in the flesh, look to the day when Miniss Kitigan will once again be Home.

Garden Island Church.

RAMBLER

HAMROCK
BAR
MODEL
Beaver Island St. James, Mich.

TELEPHONE

IT HAPPENED ON BEAVER ISLAND

by Margaret Hanley

(These are excerpts from Mrs. Hanley's newspaper column, "It happened on Beaver Island," which appeared in the Charlevoix Courier between 1947 and 1951.)

About this time of year, predictions are in order. Over here on the Island, Ole Timer says — "There's going to be a light winter. There's not a single nut on the beechnut trees." It seems as though the absence of nuts means the squirrels and chipmunks will be able to scratch and find sufficient food during the winter months.

(October 11, 1946)

With just enough snow for good sleighing since Christmas, the rising temperature towards the end of the week cleared the roads and took away practically all the snow.

The high school students, anticipating a sleigh ride out to Don Burke's Friday afternoon were somewhat disappointed. The lack of snow made the going hard for the teams drawing the sleighs so it turned out to be a hike for most of the crowd.

Saturday, an anxious fisherman skidded a fish shanty out on the ice only to retrace his steps in short order.

The sleek-as-glass ice that formed in the harbor has afforded a lot of fancy figure skating. Saturday a number of trusting skaters were down on the rink, but a good half dozen crashed through for a partial ducking.

(January 19, 1949)

Snow forts stand high and mighty in some of our front yards. Behind these forts are brave gun totin' lads and lassies fighting for a cause. This fightin' fever was inspired by the movie thriller "Custer's Last Stand," which was shown serially on Wednesday nights for the past two months or more. Mothers are wondering how long this Custer epidemic will last.

(February 11, 1948)

Icy plunges are common on the Island. This happened Saturday. In order to patch a hole in James (the young) Gallagher's fish boat, the motor was being removed and lifted to the dock. James father, James H. Gallagher and Raymond Lewis were doing the lifting and were standing on the ice. The ice broke and shot from under the two men and both went into the lake. Gallagher saved himself from going completely under by getting a hold of the dock. Lewis went into the water and under the ice. Gallagher gave assistance to Lewis and pulled the water soaked man to safety.

(April 7, 1948)

At long last the harbor of St. James is practically cleared of ice. The first of the week, the situation did not look very promising. The east wind brought down ice floes from the Straits to jam the harbor. Wednesday, the boat was unable to cross due to the stiff wind, but by Thursday, the lake calmed and much of the ice had floated out.

(May 4, 1950)

Spring is trying hard to assert itself but Ole Man Winter's batting average is striking hard with snow flurries and chilling winds. Wood piles are vanishing and as one Islander said recently, they are burning more fuel now than in the dead of winter. We're rootin' for "Spring" and by next week, we hope to report a complete knock-out, drag-out, for our warm and much preferred favorite.

(May 4, 1950)

Monday morning about nine o'clock, the drone of a motor over head sent Beaver Islanders to the windows to see who was approaching at such an early hour. Much to our surprise, there was a helicopter flying low. It was the first time many of us had ever seen one, including your B.I. reporter. It was queer looking and we strained our necks for a good look. The appearance of this strange flying machine had us wondering, but it took Ada Martin to draw conclusions.

Her youngsters came into the house to tell Mama to come see the funny airplane. One look and Ada exclaims, "My heavens, it's the Russians" and with that she hustled them down into the basement to await the worse. "But Mom, what about Daddy?" fretted one of the children as he scrambled for the stairs. "Never mind your Daddy, just get down in the basement." And there Ada and the children huddled together

. . . expecting to be blown to bits at any moment. They finally emerged from the underground shelter, a bit shaky, but mighty relieved when they found out it was the helicopter from off the Sundew checking the ice situation around the Island.

(April 12, 1951)

The State dentist, Dr. Earl Faber, flew to the Island during the week to make arrangements to return Sept. 27. Dental care will be given to pre-school and will be given to grade children. Dr. Faber and his assistant plan to stay four weeks on the Island. They will set up shop in the Parish Hall kitchen.

(September 22, 1948)

Beaver Island's present generating electrical unit, a 50 kilowatt unit, is to be replaced next spring or early summer by a 100 kilowatt unit. The unit now serving the Island has become inadequate due to the increase in electrical equipment in the homes and increased membership in the REA Co-op.

(December 15, 1948)

Dr. Raven I. McDavid, Jr., of the University of Michigan and Prof. Williamd Mackey of Lavan University, Quebec, Canada, spent several days here investigating the speech of Beaver Island as part of the survey of the speech of the Great Lakes basin, which is being sponsored by the University of Michigan and directed by Albert H. Marckwardt, Prof. of English at the university.

This survey uses the everyday drama vocabulary and pronunciation of the people of various locations as a means of discovering contributions made by various groups who have settled the old Northwest Territory. It includes six states, Michigan, Wisconsin, Minnesota, Illinois, Indiana and Ohio.

In each of these states twenty-five communities are being investigated with representatives of the older and middle aged generations being interviewed in each.

Beaver Island was chosen because it is the largest of the Island communities off the Michigan coast settled early and the type of population hasn't changed much for the past ninety years. In Island communities of this kind one can expect pronounciations and words to survive long after they have been lost on the mainland. For instance the older Beaver Islanders call a dragon fly a "Mosquito Hawk." Although elsewhere in Michigan it is known as "darning needle." "Sook, sook, sook," as a call to calves is likewise typical of the speech of Beaver Island. The researchers also found that the pronunciation of the older generation of Islanders had an interesting Keltin base. For example, the sh as for s in sparking; t as for d in island, and many other features which made their stay on Beaver Island interesting and profitable.

(June 23, 1948)

Thursday, 30 Christian Brothers arrived aboard the North Shore for a retreat at the Brother Domnan Lodge. The Brothers who comprise this group have come from New York, Philadelphia, Baltimore, Texas, Mexico, Canada and Rome, Italy. They will remain until August 20.

(July 7, 1948)

Buck of the month club announce winners!

Lee Cuddington, Vernon LaFreniere and Erwin Belfy made up a threesome the opening day of deer season. Belfy was planted in a spot while the other two went farther into the woods. Along came a big buck within close range of Belfy and two shots from his rusty, pardon, I mean trusty shootin' iron blasted at the critter. Scared and deeply hurt, the poor deer flashed out of sight.

Determined to have venison for Sunday dinner, the three began to track down the animal. At long last their quest ended. To their utter amazement, the coyotes had beat them to the draw, they had eaten the entire carcass leaving only the head. "Oh, shucks," the three were quoted as saying. The trio are now in line for the booby prize offered by the Buck of the Month Club, their names to be inscribed on an empty nail keg to be retained and used for kindling wood in case of need.

(November 19, 1947)

About the liveliest spot in northern Michigan this past week was Beaver Island.

With the yachts in from the Mackinac races and "Mayor" Bill Whalen taking over the reins of office and running the Island, the place fairly rocked.

There were 26 yachts from the Chicago Yacht Club anchored in and around the village docks.

This year, two entertainers from Chicago accompanied the yachtsmen — the Prendergast brothers. The pair with master of ceremonies Whalen put on a floor show. Community singing was a feature of the evening.

With the yachtsmen becoming more and more sold on the Island, their stay lengthens each year. A number remained over until Sunday with the floor show running three nights straight at the Tavern.

(July 27, 1949)

The Parish Hall, transformed into a night club for the occasion, took on the full effects of a swank place to dine and dance. Individual tables for four centered with burning tapers were placed in semi-circle arrangement in front of the stage where the best floor show in town was the main attraction.

Rolland McCann, acting as Master of Ceremonies, introduced "Carmen and Yolanda," ballroom dancers, who were none other than Charlotte McDonough and Lester Gallagher. "Singing Troubadours, Molly and Mike," otherwise Mr. and Mrs. Jewell Gillespie were next on the program. From the deep south came Glendora and Soapstone Johnson, a couple of black clouds. Mrs. Margaret Hanley and Richard O'Donnell presented this number, a skit and dance act.

"A Summer Day in the Park" brought down the house. The scene opened with Mary Malloy, a cute little nurse sitting in the park and policeman Johnny Andy Gallagher, being very friendly. Archie LaFreniere, Jr. repeating a command performance as a flirty gal, was the other attraction to policeman Gallagher. "Wolf" Bert McDonough did his best to become acquainted with "glamor gal" LaFreniere, under protest of the cop. In the chase — the cop after the "wolf" — the parks bathhouse door was opened, revealing "Flora Dora" Lawrence McDonough in old fashioned camisole and lace trimmed pantaloons.

Father Giles had a little trouble over a hat which brought on the song. "I Had a Hat" sung by Father Giles, Art Miller and Frank Nackerman.

Coming as a complete surprise, "Madam Human Shike," coloratura soprano dressed in an off the shoulder, long flowing gown, wide black hat tilted to an angle showing red curls, gave out with "The Kiss" from IlBacca. Folks it was Fred Spalding.

Mrs. Fred Spalding, Mrs. Rolland McCann, Mrs. James W. Gallagher, Walter Wojan, Ed Boyle, Mary Malloy and Walter McCauley appeared in the skit "A Noble Profession" showing the trials and tribulations of Dentist Wojan.

The popular beautiful song hit "Galway Bay" was sung by Father Giles.

As a grand finale, the entire cast fell in line for a grand march off the stage and down the length of the hall.

During intermission, Gene Connaghan, Buddy McDonough, Gene Burke and Donald Cole served refreshments.

Playing to a packed house, Spalding's theatrical troupe presented a full evening's entertainment Tuesday in the Parish Hall.

With the orchestra playing "The Wearing of the Green," the curtain rose for "Mrs. Murphy's Chowder" a one act play based on the song "Who Threw the Overalls in Mrs. Murphy's Chowder." The cast included:

Mrs. Nora Murphy, Mrs. Fred Spalding; Nellie Driscol, Mrs. Rolland McCann; Barney Murphy, Lloyd McDonough; Bridget Murphy, Mary Malloy; Dennis Trask, Walter McCauley; Mrs. Isaac Cohen, Mrs. Margaret Hanley; Erin Murphy, Patricia Ramsey; Helen Pearson, Mrs. James W. Gallagher.

The entire play smacked of Irish wit and good humor with a little Irish temper thrown in.

Following the play, an old fashioned song "When I Wore My Daddy's Brown Derby and You Wore Your Mother's Blue Gown" was sung by Mrs. Teddy Wojan and Johnny Andy Gallagher. The years turn back and out stepped little Colleen Nackerman and Alvin LeFraniere to sing and enact the same song. These two youngsters will be long remembered for their fine performance.

From far away places came dancers of different nations who thrilled the audience with their dance interpretations.

As a grand finale, the entire cast appeared on the stage and with the audience, sang one Irish song after another.

Announcing all the acts were little Kathleen Gatliff and Sally Martin, dressed as Leprechauns.

Rita and Jewell Gillespie about 1950.

Jewell Gillespie and Edward Palmer, 1978.

Mrs. Walter Wojan, Fred Spalding and Jewell Gillespie furnished the music for the entire performance.

(March 23, 1949)

Buddy McDonough writes from down in Texas that he is beginning to like life in an army camp. Bud says his tent mates are all from Michigan and all are Irish — which should make him feel right at home.

Buddy's sister, Joy, left Wednesday for Grand Rapids to enter training at Mercy Central School of Nursing. Joy's parents, Mr. and Mrs. Lloyd McDonough, accompanied her to Grand Rapids and will remain there for a week or more to purchase stock for the fall and winter.

(October 5, 1950)

Island talent blossomed out Saturday evening, when an entertainment in vaudeville style was given in the Parish Hall under the direction of Mrs. Fred Spalding.

Starting promptly at 8, the curtains parted on a darkened stage. Two black-out shorts were presented by Mrs. Spalding, Johnny Andy Gallagher, Mrs. Rolland McCann and Fred Spalding.

A Romeo and Juliet act in modern lingo was presented by Mary Malloy and Donald McDonough.

Two coon jiggers (black faced jointed wooden dolls) manipulated by Mrs. Spalding and Mrs. McCann gave an exhibition of soft shoe dancing.

"What Happens on a Street Car" was a comedy performed entirely in pantomine. Mr. Spalding in motorman's cap, false nose and mustache, took the part of the motorman, Johnny Andy Gallagher was the absent minded sandwich eating passenger who forgot to pay his fare. Lawrence McDonough, a "Mother with babe in arms" had an awful time trying to find car fare in a purse stuffed with baby's necessities and other paraphernalia. Archie LaFreniere, a dizzie blond, almost caused the motorman to have a wreck, he was so dazzling. Burt McDonough, a slightly inebriated gentleman managed to get aboard. Elston Pischner was the man who just made the car.

Next followed "The Little Lonely Petunia In An Onion Patch," Mrs. Margaret Hanley took the part of the lonely Petunia; Mrs. Arthur Miller, Mrs. Ethel McCann, Mrs. Edwin

Vera McDonough, later Mrs. Walter Wojan

McCann and Mrs. Spalding were the onions.

"Dem Back Wood Folks," a comedy skit in mountaineer fashion was presented by Mrs. Margaret Hanley as "Maw," Fred Spalding as "Paw" and Elston Pischner as Lukie.

An act of dancing started with Madonna McDonough and Lester Gallagher waltzing to "The Band Played On." Mrs. Bob McDonough and Lester Gallagher gave an exhibition of step-dancing. Kathy McCann as the little girl and Mary Catherine Hanley as the little boy, came in dancing to the Skirt Dance. Maw and Lukie performed some fancy capers to Good Old Mountain Music.

Preceding and during the floor show, the Beaver Island Trio, Mrs. Walter Wojan, Fred Spalding and Jewell Gillespie, entertained the patrons with a variety of good music.

Mrs. Fred Spalding directed and produced the floor show. Members of the Association of Commerce were in charge of tables and refreshments.

(February 23, 1949)

One of Earle Boyle's biscuits was displayed on an antler of the big Moosehead in the Tavern the other night. Earle's hospitality prompted him to send his friend, Bob Peterson, a sample of his culinary efforts. Peterson, discovering he could neither bite nor break the biscuits, decided they would be better as decorations. An auger was used to pierce a hole in the rock-like substance before it could be hung on the antler. Although Moosehead objected, objections were overruled.

(July 23, 1947)

The base ball bug of the Tiger variety has bitten a bunch of Island boys. To remedy the infection caused by the bite, Buddy and Bruce McDonough, Donald McDonough, Jack Cull, Donald O'Donnell and Mark Cross took themselves to Detroit to see their favorite team play.

(July 7, 1948)

Doctor and Mrs. Russell Palmer were passengers on the same plane with the McDonoughs Wednesday. They are taking their fall vacation to the Canadian Soo and other points in Canada.

(October 5, 1950)

One of the top-notch entertainers on the Island is Patrick Bonner. Pat says he never had a lesson in his life and that he started fiddling when he was a lad of 10 years. Pat's repertoire of old time tunes has no end. Besides playing the fiddle, Pat can recite poems a mile long and these poems he said he learned about forty years ago. Between his recitations and his fiddling, Pat has entertained many a gathering on the Island. Pat can't recall the names of all his tunes but knows them by the names of Island folks who stepped danced to special numbers, like the "Francis Roddy Waltz" or the "Peter Gallagher Two-Step" etc. The chin rest on Pat's fiddle is worn thin from use over the past forty years — it was given to him by a relative those many years ago. It's the proof of the use.

(October 5, 1950)

Captain Andrew Roddy.

SONGS OF BEAVER ISLAND

by Glen A. Hendrix

From Mormon times until the 1940's the people of Beaver Island made their livings by fishing, farming, sailing, and lumbering. These trades were simply those available at that time, and the careers Islanders chose reflect their energy and opportunism, but not their personalities. However, in their choice of entertainment, they had more freedom of expression than in their choice of a trade. We can, therefore, learn much of the Islanders' attitudes, personalities, and their historical background by studying their ways of entertaining themselves and perhaps we can be entertained ourselves.

Music and dance were the most important forms of entertainment on Beaver Island from the Mormon exodus until about 1940. The people of Beaver Island loved to sing! They sang as they worked on the sailing vessels; they sang at the lumber camps and at home, and they held all night singing contests in which a song was never repeated. Many singers needed to hear a song only once in order to memorize it. Captain Roddy, perhaps the greatest Island singer, "could sing any damn song that there ever was," and Johnnie Green, who claimed to be one of the Island's worst singers, knew over two hundred songs. Singing was extremely important to these people, and so it has an important place in the Island's history.

Islanders composed many songs of their own — some of these are included in this article. Many Islanders are remembered for their song writing: Dan Malloy, Frank "Conn" McCauley, Neil Deary, and Charlie "Anthony" O'Donnell to name a few. Barney "Briney" Martin made up songs from the news as he rode on horseback from St. James to Sand Bay. Most Island compositions have certainly been forgotten, but those that remain are interesting, amusing, and sometimes quite beautiful. Of course, Island compositions were only a small part of the Island's song repertoire.

The number and kinds of songs once sung on the Island is very large. Most originated on the British Isles, but many songs of American origin about sailing and lumbering were popular, as were songs written on the Island. The backgrounds of the Island's settlers, the trades they plied, the relative isolation of Beaver Island, and other unique characters of the Island help explain the popularity of singing, and the kinds of songs that were most enjoyed.

Most of the early immigrants to Beaver Island were from Ireland, and the Irish have always been noted for their love of song. This love of song they brought to Beaver Island, along with many songs of Irish, Scottish, and English origin. Many of the immigrants were from the islands of Rutland and Aranmore. Nostalgia for a common homeland (and perhaps pity for Ireland's political troubles) probably contributed to the singing tradition on Beaver Island. The wintertime isolation required that the Islanders provide their own entertainment — just as they had in their former homes in Ireland. Many of them shared fishing as a trade; they also shared songs and musical interests as well. Thus the isolation of a group of people with similar backgrounds and occupations, and an already well developed singing tradition, made song a natural mode of expression. Tradition included creating new tunes and lyrics, particularly to commemorate an event or person.

The wintertime isolation of Beaver Island did not exclude other musical traditions. During the summer the Island was an important stopping point for lake vessels of all kinds. The sailors from these ships brought many sailing songs and other songs of American origin to Beaver Island. Islanders operated their own vessels too, and regularly visited other ports where they must have learned all sorts of songs. Thus sailing songs such as "The 'Persian's' Crew," "Red Iron Ore," and "The Steam Tug 'Olsen' " were introduced to Island singers. And Islanders contributed songs of their own to the sailing songs of the Great Lakes.

Many songs from the lumbering tradition such as "The Foreman Young Monroe," "Harry Bale," and "The Falling of the Pine" were popular on Beaver Island. These songs could have been brought to the Island by men who came to fell the Island's timber, or by Islanders who worked at mainland lumber camps. It is interesting that both the sailing and lumbering musical traditions were fostered in trades plied in remote areas. Music was an important source of entertainment to sailors and lumberjacks for some of the same reasons it was important to the people of Beaver Island.

There are other reasons that can account for Beaver Island's rich musical heritage. Many of the early Islanders could not read or write. Songs were important to them as a way of remembering; many of the songs they sang, such as "The Beaver Island Girls" contain much information about people and events. Beaver Island was, and still is, blessed with a large number of fine musicians. Some of these have been mentioned, but there were many others. Mary Ellen Roddy and Katcheline O'Donnell brought many songs from Ireland. Edgar O'Donnell and Neil Deary composed many songs, but they seem to have all been forgotten. Frank McCauley wrote five of the songs in this collection, and he probably wrote others. In fact, just about everyone sang and composed songs! In this century, the "summer people" brought some new songs to the Island, and also wrote songs such as "Donegal Bay," or "Goodbye My Beaver Island Baby" (I have not found the words to this song). These songs are generally intended for group singing and are set to tunes everyone already knows.

One other factor seems to have greatly influenced the Island's musical tradition: the tragedies in which Islanders lost their lives. Ten of the songs in this collection are about shipwrecks and the untimely death of Islanders. Tragedy and death are unfortunate stimuli to song writing, but the fact that they were is a tribute to the hardy spirits of Island people. Songs of sailing, lumbering, vacationing, and some of more general American origin were added to those the settlers brought from Ireland. An unusual number of good musicians, and an unfortunate number of shipwrecks combined to produce a musical heritage as rich as the fishing grounds around Beaver, and unique to the Island.

The songs the Islanders wrote give us something no ordinary history book ever can. The songs about shipwrecks tell us not just the story of the wreck, but of the sorrow felt by Islanders who had lost friends. Often the question is asked or implied, as in "The Fate of the 'Marold II':"

Could not death wait? Could not have Fate
Passed Bruce McDonough by?
A Captain fine at twenty-nine,
He was too young to die.

Typically this question remains unanswered, and the song recognizes the will of God. These songs reveal the strong religious faith of the Islanders.' Other Island compositions are of a more cheerful nature and reveal a healthy sense of humor. Death and disaster were not the only inspiration to the Island's song writers; such was their love of song they wrote about almost anything, boat races, the death of a horse, love, and the character of other Islanders!

There are always difficulties involved with publishing folk songs. By their nature, these songs are constantly changing, so it is impossible to identify one version of a song as correct. Both the words and tune used for a song may vary, and in some cases a singer would change the tune as he or she sang. For this reason the tunes given here are only approximations of the way a singer might sing a song. I have provided tunes for the songs (or indicated a well known tune to be used) whenever possible, but many songs are preserved only in written form. Also provided is as much information as I could find regarding authorship, dates of composition, and the people and events the songs describe. The songs are in approximate chronological order. There are few rules for singing these songs, but most were meant for an unaccompanied singer, and, in the older songs, the last few words were usually spoken.

I hope this collection stimulates interest in — and memories of — Beaver Island's music. I am sure there are many other songs or poems scribbled down in forgotten notebooks or lost in someone's memory that could be added to it. There are other aspects of the Island's music not considered here, such as fiddle music, and songs of the Mormons and American Indians. Also, the Island's musical tradition is not totally a thing of the past; songs are being composed and sung on the Island to this day.

The Gallant Old Mare

Vesty McDonough and Red Anthony Gill lived on their farms on Sand Bay. Vinegar Hill is the hill on McCauley Road just before it meets East Shore Drive. Jeff Davis was the president of the Confederate States during the Civil War. This song makes fun of Vesty, who had a reputation for being very stingy. "The Gallant Old Mare" was sung to the tune of "Sweet Betsy From Pike."

The Gallant Old Mare

As Vesty was riding up Vinegar Hill,
He overtook Red Anthony Gill.
Said Gill then to Vesty, "It's your horse I might share,
Give me a ride on your Gallant Old Mare."

This Gallant Old Mare, she came from afar;
She carried Jeff Davis through Battle and War.
A friend of the Republicans, Democrats beware,
She voted for Lincoln, this Gallant Old Mare.

Said Vesty to Gill, "My horse is quite lame,
And to tell you the truth, she's not very tame.
I gladly would ride you, but it's yourself she might kill,
So I'll bid you 'good morning' Red Anthony Gill."

Granny's Song

Kitty Gallagher learned this song in Ireland; she may have written it herself. Her grandson, Owen "Conn" McCauley translated it from Gaelic. Kitty was a widow 64 years old when she left Aranmore in 1866. She came directly to Beaver Island, where she had many relatives, including eleven grandchildren she had never met. She died in 1905 at the age of 103, and there were five generations of her family present at her funeral.

Granny's Song
by Catherine "Kitty" Gallagher

Oh beautiful earth, the land of my birth
'Tis with sadness I leave you behind;
But I hope and I pray that on some future day,
That I will return to my kind.

I have nothing to fear, I go to friends that are dear,
Though long they have been gone from my sight.
They gladly will meet me, and fondly will greet me,
Where there always reigns joy and delight.

But if I will die, as the days will go by,
I hope that my soul will take flight,
O'er the beautiful earth, to the land of my birth,
Where I spent the best years of my life.

Lost on Lake Michigan and The Gallant Tommy Boyle

Johnny Gallagher, his brother Owen, and Tommy Boyle drowned in Lake Michigan in 1873 while sailing from Traverse City to Beaver Island. "Lost on Lake Michigan" is the story of their last voyage and "The Gallant Tommy Boyle" is a lament for the death of Tommy Boyle. Both songs were composed in 1874 by Dan Malloy. There are said to be two other songs about Tommy Boyle, by Father Gallagher and Mike "Mahal" O'Donnell, but I have been unable to locate complete versions of these.

Johnny and Owen Gallagher were the sons of Seamus (or James) Gallagher and Mary Gillespie. Their boat was the twenty-four foot "Lookout," built by Van Riper, keeper of the Harbor Light. The boat had a bad reputation and it had been predicted she would "put sand in somebody's ears." Late in October Johnny and Owen decided to sail to Traverse City to get supplies for Johnny's wedding.

Tommy Boyle, born in 1849, was the son of Mike Boyle, after whom Boyle's Beach is named. He had an older brother John (or Jack in the song), and several younger sisters. A few years earlier, Tommy had saved his uncle Hughie Boyle, a Carmody, and a Sullivan from drowning after their boat capsized while they were driving net stakes. The line in the song: "he saved the crew and his uncle too," refers to this incident. (Three Martins drowned in this accident when they tried to swim to shore, but the others survived by clinging to the overturned boat until Tommy rescued them.)

This is the story of the tragedy as told by Dominick Gallagher in 1938:

> Three men went out of this harbor in a small boat to go to Traverse City for supplies, and they left there in a gale of wind. They only had a twenty-four foot boat and they were all lost.
>
> That was in '73, and I was born in '67. The way I remember, my father left home with those that was drowned, and when he got to Traverse City and was ready to come back, old Captain Roddy, who had a little sailing vessel there, coaxed him to stay over and come

home with him next day when it would be comfortable. He knew it wasn't fit for them to go out in that open, that small boat, understand? It was blowing a gale of wind, it was, blowing the tops right off the seas.

My father was going right down in the boat, and Roddy said "Dominick, you aren't crazy, are you, to go in that boat today?" "Well," he said, "I'll tell you: my wife is sick, and I want to get home." "Well," he said, "it's better for your wife to be without you for two or three days than to be without you forever." And my uncle Roddy went down into the boat, and he took my father by the shoulder and he kept him from going.

We heard the next morning that the boat was lost. Well, my mother knew that my father was in the boat, you see, because he had left the harbor and went to Traverse City with them, and she didn't know of this Roddy being in Traverse City. And when the news came and the report was that all hands was lost, I remember running and hanging around mother. I couldn't realize what they were all crying about. I had six sisters and they were all home and they were all crying, too. That night they had a wake and all just as though he was there, and all the next day the neighbors came around.

Well, when this Captain Roddy came home the next day in his vessel and when they come to St. James, the harbor, they heard there that we had held a wake over father that night. My father and this Captain Roddy was great friends, and some of 'em got a jug of whiskey and they started home rejoicing that he didn't come in the boat that was lost. When my father come home he started to dance — he was always for singing and dancing when he had drinks. He never drank much except occasionally . . . But this is the way old Dan Malloy's song of it goes.

Lost on Lake Michigan
by Dan Malloy

Oh come brother sailors, I hope you'll draw nigh,
For to hear the sad news, it will cause you to cry.
Of the noble Johnny Gallagher, who sailed to and fro,
He was lost on Lake Michigan where the stormy winds blow.

Oh Johnny my dear son, in the dead of the night,
I woke from my dream, it gave me great fright.
And to Traverse City I beseech you not to go,
For you'll ne'er cross Lake Michigan where the stormy winds blow.

Oh mother dear mother, your dreams are not true,
And I will shortly return and prove it to you.
For the Lord will protect me where e'er I go,
And I'll cross o'er Lake Michigan where the stormy winds blow.

Oh Nancy lovely Nancy don't stop me my dear,
I will shortly return so come dry up your tears.
And home in our cottage full of bumpers will flow,
And I'll cross o'er Lake Michigan where the stormy winds blow.

It was in October of seventy-three,
We left Beaver Harbor and had a calm sea.
Bound for Traverse City, our destination to go,
We were crossing Lake Michigan where the stormy winds blow.

We left Traverse City at nine the next day,
And down to Elk Rapids we then bore away.
We put in our stores, and to sea we did go,
For to cross o'er Lake Michigan where the stormy winds blow.

As the day then wore on we were well under way,
And had taken our last sight of Grand Traverse Bay.
We carried all sail, and the "Lookout" did go,
We were crossing Lake Michigan where the stormy winds blow.

At nine that same night a light we did spy,
That is Beaver Island, we are drawing nigh.
With the wind from the north-west, oh how it does blow,
And we're crossing Lake Michigan where the stormy winds blow.

Oh Johnny got up and he spoke to the crew,
He said "Now my brave boys, be steady and true.
Stand by your fore halyards, let your main halyards go,
There's a squall on Lake Michigan where the stormy winds blow."

Said Owen "brother Johnny it grieves my heart sore,
To think that we'll ne'er return to the shore.
God help our poor parents, how their tears down will flow,
For we'll sleep in Lake Michigan where the stormy winds blow."

So come brother sailors, let us all shake our hands,
As we know in our hearts now, we'll ne'er reach the land.
May the great God of glory unto us mercy show,
For we'll sleep in Lake Michigan where the stormy winds blow.

The "Lookout" is running before a hard gale,
Her rudder unshipped and overboard went her sail.
And the billows came foaming like mountains of snow,
They were lost in Lake Michigan where the stormy winds blow.

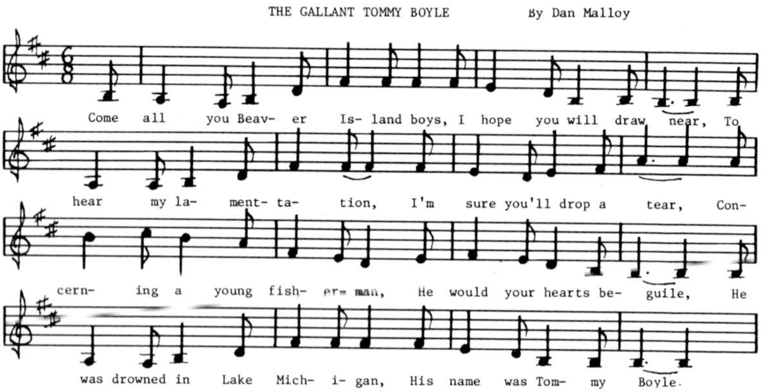

The Gallant Tommy Boyle
by Dan Malloy

Come all you Beaver Island boys,
I hope you will draw near,
To hear my lamentation,
I'm sure you'll drop a tear.
Concerning a young fisherman,
He would your hearts beguile,
He was drowned in Lake Michigan,
His name was Tommy Boyle.

When his sister heard the news,
She dropped a silent tear.
She wrung her hands in anguish,
Crying "Oh my brother dear!
What shall I do these winter nights,
When you would on me smile?
For my heart is nearly broken,
For my brother, Tommy Boyle."

When his father heard the news,
Distracted he did run,
Crying "Neighbors dear, what shall I do?
I've lost my darling son!
Like a pilgrim I will wander,
And I'll travel many a mile,
And all my lifetime I will mourn
For my own son Tommy Boyle."

The Reverend Father Gallagher,
Great praise to him is due,
He reconciled his father,
And preached the Gospel true.
He prayed for his salvation,
That the angels on him smile,
May his soul find rest in Paradise,
That gallant Tommy Boyle.

He was proper, tall, and comely,
His age was twenty-three;
He was as fine a young man
As you could wish to see.
He was proper, tall, and kind to all,
And on his face a smile,
And I hope he's well rewarded now,
That gallant Tommy Boyle.

He was a gallant boatman,
And fishing he did know;
And when they were imperiled,
To release them he did go.
He saved the crew and his uncle too,
And Heaven did on him smile,
And I hope he's well rewarded now,
That gallant Tommy Boyle.

Now to conclude and finish,
Dear Jack, you'll understand;
I'm going to meet my mother dear,
She's in that blessed land.
She was always kind and true to me,
And I think it but little toil,
To sing her praise, God rest her soul,
Your mother, Tommy Boyle.

The U.S. "Lightship 98"

The "L.V. 98" was a lightship which was stationed over the Lansing Shoals. She was built in Muskegon in 1918. In 1922 a lighthouse was built on the shoals and the "L.V. 98" was sent to the Atlantic. Frank McCauley, author of this song, worked on the "L.V. 98" as a cook.

The U.S. "Lightship 98"
by Frank McCauley

They may boast of their dreadnaughts and cruisers likewise,
Sub-chasers and airplanes that soar in the skies;
But there is another whose fame I'll relate,
It's Uncle Sam's little watchdog, the "L.V. 98."

In the lanes of our shipping a boulder does lie,
Which nature had formed as the ages rolled by;
And beneath the Lake's surface in silence does lay,
Like a beast in the jungle in wait for its prey.

Here Uncle Sam's watchdog at anchor does ride,
And swing to and fro with the winds and the tide;
But she constantly guards while the billows do roll,
O'er that freak of dame nature, the famed Lansing Shoal.

Her watch room is white and her hull is bright red,
And a background of black o'er her maindeck is spread.
Her tall stately tower is looked on with pride,
For it warns all mariners this rock to avoid.

Her deep-throated siren how loud it does roar,
When the fogs and the mists envelope our shore;
Her miniature beacon, it shines out so bright,
As it sends forth its rays and illumines the night.

Two skilled engineers to her engines attend,
And to keep them in order, their energies bend.
The cook and the seaman, the captain and mate,
Comprise the whole crew of the "L.V. 98."

Sometimes we are lonely and feel indisposed,
For we are estranged from our friends and our foes;
But there's one earthly blessing I'm sure we enjoy:
That no bill collector ever can us annoy.

In the watches of night not a sound can we hear,
But the gulls and caw-weens that bring us some cheer.
With their wrangling and clamor they bring some respite,
As they sound their wierd notes in the stillness of night.

Now my seafaring comrades, wherever you roam,
Remember this watchdog has guided you home;
When you're safely landed do not hesitate,
To take off your hats to the "L.V. 98."

W.P.A. Song

The present East Shore Drive was constructed in 1937 by the Work Projects Administration. Some of the workers on the East Shore Drive Project were Charley "Anthony" O'Donnell, Anthony Green, Danny Green, and Willie Ricksger. Gus Meilke, owner of the sawmill in St. James, was their boss. The story is that O'Donnell, who wrote the song, was hard working, but the others tried to do as little as possible. Tom Gatliff drove the snowplow. The "Escanaba" was a coast guard icebreaker that brought hay to the Island one winter when supplies ran low.

W.P.A. Song
by Charlie "Anthony" O'Donnell

Sloptown is a darn good road, so is the King's Highway,
But, of course, it's got nothing on the road up to Sand Bay.
It was engineered by Mielke, he and his gallant crew,
Of course, advised by Gatliff, who told them what to do.

When winter spreads its mantle, and Tom takes out the plow,
It's thrilling for to watch them with the snow about her bow,
As he comes thundering up the road while widely drives the spray,
Just like the "Escanaba" when she comes in with the hay.

We are now quite well represented by the wise men from the east,
They strive with one another to see who can do the least.
Their average height was five foot eight, but that I cannot say,
For I never saw them standing up on the W.P.A.!

Donegal Bay

The Allen family wrote these words to the tune of "On Top of Old Smokey" about 25 years ago. Henry Allen owned Donegal Bay at that time. "Allen's Cabins" were the first buildings on the Bay and the words were written to sing around the beach fires in the evening, often with visitors who were staying in the cabins. The Blarney Stone referred to in the song is a large boulder near the middle of the Bay, which, at the time the song was written, sat a short way off the shore and required a wade to reach. Lately the stone has been submerged, due to the high lake level. The song is still sung as it was written: around a fire on the beaches of Donegal Bay.

Donegal Bay
by The Allens

If you want to have fun, friends,
Away from it all,
Just pack up your things,
Head for old Donegal.

There's lots of white sand dunes,
Lake Michigan blue,
And mountain to climb high,
With a magnificent view.

On top of Mount Pisgah,
Above the tall pines,
The scene far below you,
Is worth the great climb.

You'll find a warm sand spot,
And spread your mat gay,
Relax in the sunshine,
The rest of the day.

On the shore waves are breaking,
Against a large rock,
It's known as the grandson,
Of the Blarney Stone stock.

You'll wade out to kiss it,
In the true Irish way,
You'll always remember,
Old Donegal Bay.

You'll reluctantly leave there,
And you will all say:
"We'll return every year, friends,
To Donegal Bay."

Donegal Bay.

Turner's Pure Bred Bull

At one time it was common for Island farmers to share the use of a bull. Turner's bull was owned by Dan Turner, Harlem Gallagher, and Pat Bonner. They are the stockholders mentioned in the song. "Young James" was James Bowery. Clara Schmidt and Charley "Anthony" O'Donnell wrote this song about 1940. Mrs. Schmidt, on a recording of her song, uses two distinct melodies and a number of variations of each, making it impossible to accurately write out the music. The melody presented here is one that Mrs. Schmidt used, and it works for the whole song, with a little imagination!

TURNER'S PURE BRED BULL By Clara Schmidt and Charley O'Donnell

Turner's Pure Bred Bull

by Clara Schmidt and Charlie "Anthony" O'Donnell

Come all you jolly farmers
To whom it may concern,
And listen to my story,
It won't detain you long.
It's of an old bull animal,
So proud and beautiful,
He was known in town and country
As Turner's Pure Bred Bull.

'Twas in the month of September,
Nineteen forty was the year,
Cattle buyers from near and far
To the Beavers they did steer,
To separate the natives
From their cattle and their wool,
And they winked at one another
When they saw Turner's Bull.

When the boat arrived at St. James,
And landed at the dock,
Young James was there with his red truck,
As solid as a rock.
There was Ramsey, Schmitty, and Curry,
And to them James did say:
"Turner's got as fine a bull,
As ever was fed on hay."

Then straightway off to Turner's
Went this jolly band.
They met Turner at the gate,
And he gave them the glad-hand.
"What have you got?" says Schmitty,
"Have you any wool?"
""No, I havn't," says Turner,
"But I've got a lovely bull."

"What is the price you ask for him?"
Said Schmitty, "Eighty I will give,
I cannot pay any more,
That bull will cost me plenty
Frank Dan he sold one not so good
For manners or for looks,
He sold him last October,
And he got a hundred bucks."

Said Schmitty, "Eighty I will give,
I cannot pay any more,
That bull will cost me plenty
When I get to the main shore.
For feed and transportation
You know is very high,
If you don't take my eighty bucks
I fear I cannot buy."

Next day Jim Curry came along,
With his lovely brogue and smile,
And he met Turner at the barn,
And they talked for quite awhile.
Jim said "go and fetch the bottle,
It's over in the car,
When it's done we'll get another,
For it isn't very far."

Then back to the barn they hastened
To look the big boy o'er.
Curry said, "He's as fine a beast
As ever I saw before.
Now tie that rope my bully
To the ring that's in his snout,
And we'll have him at the Harbor
Before the boat goes out."

Now Little James came on the scene,
To lend Turner a hand.
Jack Turner, too, was on the job,
So stately, tall and grand.
He says "There's money in this, boys,
When we but get our pay."
So off to town they hastened,
They didn't want to stay.

Jim Mooney and Harlem sighed
As they saw him led away,
Pat Bonner played his violin,
And not a word did say.
And Mrs. Bradley stamped the sheet,
As she saw him led away.
In due time they arrived in town,
And landed at the dock,
The women and the Indians,
Around him they did flock.

For never a finer animal,
In St. James did they see,
At the harbor of the Island,
Well known to you and me.
When Turner got his money,
To the Shamrock he did go,
And he said, "I'll have a highball,
My sporting blood to show."

"Oh, I'll battle with the Frenchmen,
When I get another beer;
They licked them over yonder,
Why can't they do it here?"
When Turner had his liquor,
Then to them he did say,
"I'd better hasten homeward,
Or the Devil will be to pay,
To settle with the stockholders,
Before it goes too far."
Were the last words he uttered,
As he stepped into his car.

Jim, he is overcome by grief,
The Harlems they are too,
All are up town in mourning,
What else can they do?
Jack Turner he is tickled,
And to Little James did say,
"Papa's got the finest bull,
That ever was fed on hay."

The Fisherman Yankee Brown

"The Fisherman Yankee Brown" is one of many Island songs that make fun of people. "Yankee" Brown was Wesley Brown, who married Bridget Malloy. He was from New York State, but he lived awhile at St. Ignace before moving to Beaver Island. Wesley Brown is mentioned as the caller at the dance in "The Beaver Island Girls."

Lying was a respected form of entertainment on the Island and all over Michigan. One Islander is said to have used his talents to find a wife in Ireland. He told the girls there he owned fifty fish tugs and used twenty sacks of grain a day to feed his chickens!

This song is probably a parody of a lumbercamp song "Shanty Boys in the Pine" (also called "The Shanty Boys"). The tunes to "The Fisherman Yankee Brown" and "The Shanty Boys in the Pine" are almost the same, and some of the lines are the same. There are other verses to this that have been forgotten, and some verses may have been added after the song was composed; it is the sort of song that tends to be elaborated!

THE FISHERMAN YANKEE BROWN by Frank McCauley

The Fisherman Yankee Brown
by Frank McCauley

My boys, if you will listen now, I'll sing you a little song,
It's about a man you all know well, whose tales are always strong.
He is a well known fisherman who lives here in our town,
He's a very noted fisherman, and his name is Yankee Brown.

Chorus: Sing high tiddy high, high tiddy um,
With a whack fal-lor-a-lie.

He was born way down in N'York State, in a family of sixteen,
And his boyhood deeds would fill a book the greatest ever seen.
He farmed and lumbered half the state, but found it too confined,
So he came to Beaver Island in the year of seventy-nine.

Chorus

When here he took up fishing and he followed up his trade,
And several times he's told me that his fortune he has made.
He says he's acres by the thousand and pound nets by the score,
And twenty thousand dollars in a bank in Baltimore.

Chorus

His boat he calls the "Mattie," and the truth to you I'll tell,
Several times he has rebuilt her and he can sail her well.
For once when beating down the Lake he had to come about,
He heeled her in the seas so far she scooped in thirty trout.

Chorus

He swore when he was fishing once down on Hog Island Sound,
That alone he caught a white fish that weighed a hundred pounds.
At first we didn't believe him, then he began to swear,
He said he'd bring his young son out and prove the story there.

Chorus

Oh when he went to Charlevoix, to preach the gospel there,
He swore that he was anti-Christ, and for them to prepare!
His feats they are most numerous, and the truth now I must tell,
The Old Nick never will be at rest 'til he gets the Yankee in Hell!

The Beaver Island Girls

"The Beaver Island Girls" was written in 1874 by Isaac Wright, a school teacher on High Island. It is about Fourth of July celebrations on Beaver Island, which included dances and sailboat races around Garden Island. "The Beaver Island Girls" is remarkable because it includes many place and personal names. Henry and Charlie Wright lived near Sand Bay, where Henry owned a store. Fred Butts, the sheriff, lived at the Head, near the lighthouse. John Vincent was an Indian shipwright who lived on Garden Island. John Dunlevey, a cobbler, had a shop next to the old Beachcomber Bar, and Dan Williams lived in a log cabin below the hill where the church now stands. The "Uncle Mike" in the song is Mike Martin. Wesley Brown was from St. Ignace, and had married into the Malloy family. He is the subject of the song "The Fisherman Yankee Brown." Tip Miller was the first chief of the Harbor Life Saving Station, and the fourth keeper of the Beaver Head Light. He was a Mormon who stayed after the exodus. The Beaver Island Girls must have been famous dancers, for many of the names mentioned in the song seem to belong to men who did not live on the Island, but came for the dance!

The Beaver Island Girls
by Isaac Wright

The girls of Beaver Island are a very merry set,
A finer lot of damsels I think I never met.
To understand this ditty, you must know 'tis their delight,
To sail around the harbor on a pleasant summer's night.

I went to Beaver Island to see the girls and boys,
Who were talking and laughing and making lots of noise.
I asked the cause of an old man whose locks were turning white,
Said he "the Beaver Island Girls are coming out tonight."

I looked east and Captain Falcon was coming into space,
With his kites a flying, he was bound to win the race.
I said to the Captain, "what does this mean when you are flying light?"
He answered me "the Beaver Island Girls are coming out tonight."

"This is the very reason I am now coming into port,
So myself and jolly tars can have a little sport."
Then he ordered out his boats, those yachts both black and white,
To carry those Beaver Island Girls who went sailing out that night.

Charles Alloire was giving orders to fit out the "Jesse Ann,"
While running round the beach came every boy and man.
The barber he was shaving some to put them in good plight,
To go with the Beaver Island Girls who were coming out that night.

The "Fair Play" left Fox Island for this port with Captain Sam Leblau,
She beat the "Flying Dutchman" and never once did yaw.
Cried Sam "what in the deuce makes this craft run so light?"
'Twas because the Beaver Island Girls were coming out that night.

Hark, hark, what vessel's that firing a signal gun?
That is Captain Roddy, who is always full of fun.
He was bound to reach this port today, let the wind blow high or light,
To ride with the Beaver Island Girls when they came out that night.

Ben Campbell went sailing on a fine river cruise,
His "Flora" would not sail a bit until he heard the news,
But to see the foam around her bow was quite a pretty sight,
When he learned the Beaver Island Girls were coming out that night.

Hercules Lashontz, a cooper too, is a very nice young man,
That's what some folks say, let them deny it if they can.
While others stand ready to swear he is a geniune blatherskite,
Who is trying to gull the Beaver Island Girls when they come out at night.

John Vincent is a first rate hand to build a little boat,
It takes him longer to do it though than anyone afloat.
Because he is running hither and yon to get a sup and a bite,
And to ask some of those Beaver Island Girls if they'd come out tonight.

Now girls if you should want a beau, Dan Williams is just your man,
He lately came to this town from famous Cheboygan.
He is a stunning fellow, knows how to fly a kite,

And play for the Beaver Island Girls who are coming out tonight.

Charley Pausuette's from Cheboygan, a bully boy is he.
Who can spark the girls upon the land or sailing on the sea.
He is not much on style, but his word is honor bright,
So say the Beaver Island Girls who are coming out tonight.

Mr. Campbell and a friend were fishing rather late,
Those cussed little perch nibbled off most all their bait.
At length each got a most tremendous bite,
When both exclaimed "I'll bet the Beaver Island Girls are coming out tonight."

I saw a couple floating round the harbor in a very leaky boat,
The pesky thing could hardly keep those two loving souls afloat,
While he was leaning up to her in a way that was natural quite,
Whispering "do you think those Beaver Island Girls are coming out tonight?"

I called on John Dunlevy, he was pegging very fast,
And did not seem to mind at all how hard he hit his last,
Saying "those shoes must finished be and blackened up so bright,
For a jolly Beaver Island Girl who is coming out tonight."

I went down to the cooper shop, Mr. Curtis was not in,
He had just gone down to Gallagher's to get a horn of gin.
But Charles Smith spoke up "I'll get up a dance on sight,
If those handsome Beaver Island Girls are coming out tonight."

After a brief rest I left there and traveled to the Point,
With Dormer gone, I found things sadly out of joint.
Though I heard a well dressed gentleman ask the keeper of the light,
If she thought the Beaver Island Girls have all come out tonight.

We will go over to Lamontang's to have one little dance,
Where everyone who likes that sport can have an equal chance,
While Wesley Brown was calling off he bawled with all his might,
"Those lovely Beaver Island Girls have all come out tonight."

Now all you bonnie lassies 'tis time for us to part,
I wish you all success from the bottom of my heart.

May you never get worse counsel that you get from Isaac Wright,
Who thanks those Beaver Island Girls for coming out tonight.

Tim Smith was coming over from Charlevoix some cattle for to buy,
His mouth was filled with pigtail and blood was in his eye.
He called to Steve "shake out that reef, haul on those halyards tight,
For I must see the Beaver Island Girls who are coming out tonight."

When the "Black Scow" is in this port 'tis hard to get her out,
Though you will find it different when once she's come about.
She then can beat that "Goldsmith Maid" whose heels are very light,
So her crew can view the Beaver Island Girls when they sail out at night.

At last our friend John Bonner has launched his vessel from the stocks,
And we hope she earns him a pocket full of rocks.
Also we want him when he gets her all rigged up right,
He will give the Beaver Island Girls a sail some pretty night.

A passenger down on the steamer "Badger State,"
Heard Captain Clark telling his engineer and mate,
To hurry up their cokes for landing Henry Wright,
Who wished to join the Beaver Island Girls who were coming out tonight.

The "Coyhoga" too brought down a lot of girls,
Dressed out in city style with their furbelows and curls.
We were glad to see them coming, for that is all right,
For they are some of the Beaver Island Girls who are coming out tonight.

So many fish boats coming I never saw the like,
Johnnie Martin, Dan Malloy, Kelly, and Uncle Mike.
Kelly reached the dock first, rushed up to Charlie Wright,
With "see here, look here, are the Beaver Island Girls coming out tonight?"

The last boat coming in that day sailed faster than the rest,
I wondered who was in it, returning from the west.
'Twas Tip Miller and Ed Bedford, keepers of the Head Light,
Just in time to see the Beaver Island Girls when they came out that night.

Fred Butts being sheriff then had called a posse out,

He could get no one but Butler, who was troubled with the
gout.
They marched around from point to point, both spoiling for a
fight,
To protect the Beaver Island Girls who were sailing out that
night.

John Gallup is a cooper who has long been in our town,
And who in some respects takes all the others down.
You would laugh to see him flop around and hoop his barrels
tight,
When he heard the Beaver Island Girls were coming out that
night.

Old Tom Sweet

Old Tom Sweet was a huge horse owned by George Sweet, who ran a sawmill near Sand Bay during the 1880's. Old Tom was used for farm work, and to pull logs from the woods or boom them along the shore to Sweet's Mill. It was while pulling these logs that Tom "broke his collar bone." The shoemaker mentioned in the song was Johnny Gallagher. Johnnie Martin wrote the song and in it refers to himself as "Old John." Mary was Martin's wife. Red Mike Boyle and Vesty McDonough lived on farms near Sweet's Mill.

Old Tom Sweet
by Johnnie Martin

Old Tom Sweet had no shoes on his feet,
And Old Tom didn't care.
Shoemaker said "If you bring him down
I will make him a pair."

Chorus: Now Old Tom Sweet you are dead and gone,
And the crows will pick your bones.
A finer lad we never had,
But you broke your collar bone.

Red Mike did cry when Old Tom did die,
For he used to plow his land,
Nine inches deep Old Tom would sweep,
And he turned up all the sand.

Chorus

Oh, Mary was in Sand Bay,
She was sorry for Old Tom.
She ran home a'crying,
And she has told Old John.

Chorus

Now when old Vesty heard the news,
His long tooth did gaze
He said he would take Tom down,
"I will give him some hay."

The Loss of the "Clifton"

The loss of the whaleback steamer "Clifton" in Lake Huron on September 22, 1924 inspired these four songs. It is thought that her self-unloading boom broke free and swung over the side, rolling the vessel and causing her cargo of stone to shift. The combination of shifted cargo and the weight of the boom hanging over the side then capsized the "Clifton." The real cause of the loss of the "Clifton" can never be known since all hands were lost.

There were four men from Beaver Island on the "Clifton" when she was lost: Emmett Gallagher ("Emmett Salty"), Anthony McDonough, Joe Scheid, and Peter Burns. Emmett Gallagher was the captain of the "Clifton"; Anthony McDonough the second mate and Joe Scheid and Peter Burns were wheelsmen. Emmett's brother, Peter Gallagher, wrote "The Clifton Tragedy."

After the "Clifton" was first reported lost, there were other reports that she had washed ashore in some remote part of the Lake, or even that she was at Oscoda, a port she was too large to enter. Thus for several days friends and relatives of the "Clifton's" crew hoped and prayed that the vessel was safe, yet feared she was lost.

The last two lines of "The 'Clifton's' Crew" by Pat Bonner are sung to the melody of the last two lines of the other verses. There is a line missing from the words for "The Clifton."

The "Clifton" Tragedy
by Peter Kenneth Gallagher

A gray-haired mother knelt in prayer
Before the holy light
And the image of the Christ was there
To cheer her through the night.

She prayed to Him to save her boy
He, who on the Lake of Galilee
Had changed a raging tempest
To a calm and peaceful sea.

Now across the bay of treacherous Saginaw
The seas roll mountain high
The hurricane now sweeps her deck
With lightning in the sky.

The thunder roared; a deluge poured
The sky is black; 'tis night.
The Clifton struggles gamely on
Though she fights a losing fight.

The Captain's face is pale and set,
His voice is hoarse and stern
As he turns to the helmsman,
His hand on the ship's chadburn.

"Part a Point! Steady!
We'll hold her in the wind."
The helmsman answering turns her
To the order given him.

But He who ruled the waters
Held the fate of the stricken ship
Thought fit, perhaps, to take them
On that long Eternal trip.

"Christ have mercy on our souls, boys!"
Spoke the Captain to his men.
And each sailor hoarsly murmured
Clear, though lowly, the "Amen."

With her cargo fastly shifting
She is broadside to the sea
Now no human power can save her
"Christ, we trust alone to Thee."

"Pray, oh Mother Mary, for us!
Thou to whom thy Son didst say
That each favor thou wouldst ask Him
He would grant. Oh Mother, pray."

Thou of whom the theme was written
Mother, Mary, Star of the Sea
Be our Guiding Star, and Mother
Intercede, we pray to thee."

The "Clifton"

Steaming out of the Straits of Mackinac,
She blew her last salute.
Five whistles told her company's name,
Four blasts then told her owner,
As she cleared for Detroit City,
With her cargo all of stone,
"Steer south-east by east" was ordered,
"South-east by east" came the reply.

. . . . .

Then the "Clifton" squared away
The wind blew on her after quarter,
'Til abreast of Saginaw Bay,
When a terrible squall burst on her,
With a change of wind and sea,
And without a moment's warning,
She is forty miles from lee!

And her able crew of sailors,
All so young, so brave, so gay,
Little dreamed that ere the morning,
They would find a watery grave!
Leave no farewell words behind them,
Balm to hearts that soon must break,
In the springtime of their manhood,
Ah! How cruel the hand of fate.

Broken now the ties that bind them,
To this life they once held dear,
In a fairer land to find them,
In a holier atmosphere.
Never more to hear their footsteps,
Never more to clasp their hands,
'Til we are with them numbered,
In a fairer, brighter land.

The "Clifton," photographed here before her name was changed from "Samuel Mather" and a self unloader installed.

Pat Bonner

THE CLIFTON'S CREW By Pat Bonner

The "Clifton's" Crew
by Pat Bonner

We have heard of many happenings since last year first began,
With crimes and troubles caused by war from Italy to Japan,
But sorrow had not filled our hearts 'til the news it came around,
The steamer "Clifton" she was lost and all her crew were drowned.

It was on one dark September day the last was ever known,
Of the missing freighter "Clifton" that was loaded deep with stone.
With her derrick standing foreward and her hatches battened down,
She was last seen passing down the Straits, for Detroit she was bound.

It was then down on Lake Huron, she had steamed way out of sight,
For to seek her destination, and to battle with the night.
We thought not of our comrades there beneath those stormy skies,
Lake Huron's waters for to end their young and happy lives.

As the night drew on with darkening clouds, the wind would shift and change,
With a heavy sea from the south-west and a downpour of heavy rain.
The thunder in its anger crashed, as the lightning lit the deep,
Where darkness in its fury dwells, and desolation sweeps.

As time went on with rolling seas, her cargo shifted then,
And soon all hopes had vanished there among those thirty men.
The crew had all got orders from their captain and their mate,
For to wear their life belts through the night on that foundering ship of fate.

It is hard to say, we know not when that fatal hour comes,
But the waters of Lake Huron soon claimed many a mother's son.
And home today where orphans prayed and mourned their father too,
Who had lost his life that stormy night among the "Clifton's" crew.

We had hoped at first it was not true; we wished them safe from harm,

Or had seeked some peaceful shelter from the fury of the storm.
But it's where Lake Huron widens, there's no sheltered port in view,
That is where the "Clifton" she had sunk with her cargo and her crew.

Some may have thought of home and friends, some thought of God alone,
And some of their wives and children that are waiting them at home.
'Twas but a few days after when some wreckage washed ashore,
Which had told the tale of that missing ship and our friends we'll see no more.

Her captain Emmett Gallagher, a man so brave and young,
At his parents home they now must know the tidings of their son.
It is now the pangs of sorrow will pierce his mother's heart,
As she learns his doom has come so soon, her boy from her must part.

'Twas but a few short weeks before when Emmett he had said:
"There is a place out in Lake Huron that will not give up its dead.
It is called a sailor's graveyard," those words he was known to say,
As he stood out on the "Clifton's" deck, he showed them where it lay.

We had there three more friends of ours, who are in his company,
Brought up on Beaver Island and had followed up the sea;
'Twas Anthony McDonough, Joe Scheid and Peter Burns,
We hoped to meet them all again and wished their glad return.

Young Anthony, a mate on board, he has sailed the ocean blue,
He had come back to the Great Lakes and had joined the "Clifton's" crew,
To be among companions of his early boyhood days,
He was doomed like many more on board to fill a sailor's grave.

Among our four companions on the "Clifton" that were drowned,
Joe Scheid he was the only one of them 'twas ever found.
Their lifeless forms in silence now are drifting far and wide,

'Til they waste away from day to day and wear with time and tide.

Their fathers they will not forget that day so sad and long,
When their thoughts were of the "Clifton" and their dear boys that are gone.
Also their sisters they will weep and mourn their brothers dear,
Whose smiles they'll never more behold, their saddened hearts to cheer.

Our homes have not beheld the like since many years before,
Not since the steamer "Vernon" sank out off Wisconsin's shore.
As some may hold in memory still and some were then not born,
Who had perished on the "Clifton" in the darkness and the storm.

There are three of them 'tis sad to say no mother's home have they,
They are gone before their children and are long since laid away.
They say there is a Mother who has seen through many storms,
And has prayed for those who called to Her in danger and in harm.

She is our blessed Mother as the holy people say,
Is called the Queen of Heaven and had sorrows in Her day.
She has listened to many a midnight prayer from many a sailor boy,
And with God has blessed those hidden graves; they now in peace may lay.

'Tis well for us to think upon the words that God has said,
As are told from out the Scriptures, we should not forget the dead.
You may kneel at your bedside when the quiet shadows fall,
Then in your prayers remember those who were once among us all.

And ask that we may gather with the angels in their flight,
And forget the steamer "Clifton" and that dark September night.

The Seaman's Lament
by Frank McCauley

My seafaring comrades, attend to my lay,
For Death, that grim reaper, has taken away,
The fair Emmett Gallagher, so loyal and brave,
And in Lake Huron's waters has found him a grave.

On the steamer "Clifton" a captain was he,
He was genial and cheerful, kindhearted and free.
May his soul with bright angels to Heaven ascend,
For the poor and the needy, they found him a friend.

September the twentieth, nineteen twenty-four,
He took on his cargo and steamed from the shore.
Then to the eastward his course he did steer,
Little dreamt our bold hero that death was so near.

Now twilight it deepens, and far in the west,
The sunbeams have faded; all nature's at rest.
But see those dark cloudlets and swift is their flight,
And bright lurid flashes illumine the night.

Then Emmett his crew to the deck he did call,
Saying "Boys, be prepared for the worst, one and all!
My comrades, get ready your lives to defend,
For yonder dark cloud does some evil portend."

Now the tempest is raging with strength in its might,
And loud rolls the thunder, and dark is the night;
And the mad billows leap like wild beasts from their lair,
And in their wild rush not a life will they spare!

And as they roll on over that structure of steel,
The steamer does tremble from foretop to keel;
For those mighty forces that swiftly roll on,
Have torn out her vitals – all hope now is gone!

Now the steamer is sinking! O pity their plight!
O'er that wild waste of water no help is in sight.
There's none to console them, and it grieves me to say,
To Lake Huron's dark waters its toll they must pay.

God help their poor parents who shortly will mourn,
When they learn that their dear ones will never return.
And kind loving sisters in silence will weep,
For the loss of their brothers consigned to the deep.

Farewell then, my comrades, no more shall we meet.
Your sweet smiling faces no more shall we greet;
No more your lithe footsteps our pathways shall tread,
Until Huron's dark waters shall give up its dead.

The Ill-Fated "Vernon"

The year-old steamer "Vernon" foundered in a bad storm off Twin River Point in north-western Lake Michigan on October 29, 1887. The "Vernon" had steamed from Glen Haven and was on its way to Chicago, first stopping at Beaver Island to pick up six passengers: "Big Rosie" Gallagher, Bridget, Mary, and Catherine Gallagher, Charles Gallagher, and Hannah Maria Malloy. Big Rosie was going to Chicago to try to convince her husband, Cundy (not the Cundy Gallagher who owned the bar), to return to her. Bridget, Mary, and Catherine were daughters of Mike Gallagher ("Mike Mahal Ruah") and Susan Mooney. Bridget was born in 1861, Mary in 1862, and Catherine in 1863. Hannah Maria Malloy, born in 1867, was on her way to Chicago to be married. Charles Gallagher was the brother of "Salty" Gallagher, whose son Emmett was lost thirty-seven years later on the "Clifton." Hugh Connahan was to have been a passenger on the "Vernon," but at the last moment he decided not to go.

Of the sixteen passengers and twenty-five crewmen on board the "Vernon," only one man survived — Axel Stone. Stone clung to a makeshift raft all night while his companions slipped into the icy water one by one. The Steamer "Superior" passed this raft, and noticed men and women clinging to it, but did not stop to rescue them. Subsequent investigations of the shipwreck suggested Captain Thorpe of the "Vernon" had been drunk and his vessel overloaded at the time of the tragedy.

The Ill-fated *Vernon*
by Frank McCauley

All you true feeling Christians, I hope you will draw near,
And learn my doleful story, I'm sure you'll drop a tear,
Concerning our dear comrades who lately left our shore;
They left us to lament, we'll never see them more.

On October twenty-fifth the "Vernon" steamed away,
On board our six dear friends with spirits light and gay.
And little did they think as they bid their friends goodbye,
That their fate would be so soon in watery graves to lie.

It was on a Friday evening they headed up the Lake,
And no one thought the "Vernon" would be lost before daybreak.
Their peaceful sleep was broken by cries that filled the air,
The steamer she was sinking, grim death at them did stare!

It was off Two Rivers Station the "Vernon" did go down,
With five and twenty passengers, for Chicago she was bound.
Some of them manned the lifeboats, more the rafts did take,
In hopes some passing vessel would save them from the Lake.

The morning dawned in terror for that cold and drowning crew,
Adrift in the icy water, praying to be rescued.
The snow was driving down as some numbed by the cold,
Sank in the roaring Lake, their sufferings untold.

What sudden hope is this that brightens up each eye!
It is the joyful sight of a steamer drawing nigh.
Oh God! what anguished hearts as she sped swiftly on;
What doom awaited them before the setting sun!

Here I ponder their awful fate, it fills my heart with grief,
To see them cold and dying, they could get no relief.
Their strength is failing fast, in the angry Lake they die,
Far from their family and friends as they bid this world goodbye!

Oh, who was this unknown wretch that saw his fellow man,
Fighting for life in the water and would not lend a hand
To save them from this cruel fate? No pity did he feel,
This base and cruel man whose heart was hard as steel.

The "Vernon."

The "Marold II"

The tanker "J. Oswald Boyd" ran aground on Simmon's Reef on November 8, 1936, with a cargo of 900,000 gallons of gasoline. The Beaver Island Transit Company then paid $3,700 for the salvage rights of the "Boyd." On January 1, 1937, the "Marold II" left Charlevoix to salvage some of the gasoline in the "Boyd." On board the Marold were Ludlow Hill, his son Leon, Bruce McDonough, Everett Cole, and his brother Raymond. Bruce McDonough and the Coles were residents of Beaver Island. Two men, Jack Gallagher and W. H. Wallace overslept and missed the boat.

The "Marold II" exploded while pumping gas from the "Boyd," and all five men were killed. The flames of the explosion were seen forty miles away. The bodies of the Coles were found on the hulk of the "Marold," but they were so badly burned they could only be identified by their dental work. Captain Hill's body washed up on the south-west end of Beaver Island, and his son's body was found near Fairport. The body of Bruce McDonough was never found.

"The Fate of the 'Marold II' " by Vaughn Ogden has been published in the Charlevoix *Courier* and the Beaver *Beacon*. Ogden was a school teacher at the "Little Red Schoolhouse" on the King's Highway.

The Fate of the "Marold II"

by Vaughn Ogden

The fateful crew
Of the "Marold II"
Sailed out from Charlevoix Bay.

The sea was light;
No ice in sight;
Though it was on a New Year's Day.

Northward they sailed,
The land was veiled
In a haze of uncertain hue.

A sound was heard,
"On," was the word.
"To the wreck," said the doomed crew.

"To the tanker 'B'yde,'
With 'ile inside.
We'll pilfer her bounteous store."

"Our barrels will hold,
Of this liquid gold,
Ten thousand gallons or more."

Hog Island now
Was on her bow.
They had but an hour to run.

The engines strain
With might and main,
And soon the hour is won.

Tied along side
Of the tanker "B'yde"
They go to work with a will.

Some on the deck;
Some on wreck;
There are plenty of barrels to fill.

"Cigarettes out"
The leaders shout,
"We want no calamity."

"A little light
Might soon ignite
This fumous volatility."

The day soon goes.
A light wind blows.
A seagull hovers around.

Can his eye see
Catastrophe,
Where the tanker went aground?

A Charlevoix guard
Is staring hard
At a pillar of smoke and fire.

At Epoufette,
All eyes are set
On the column, mounting higher.

They see the flames
From off Saint James.
Can their sons have come to grief?

They mutely stare

In dumb despair
At the blast from Simmon's Reef.

At Mackinac,
The coast guards saw
That the sea had claimed its toll.

Charlevoix's out,
Yet there's great doubt
If they can save a single soul.

They all stand by
As the flames mount high.
They dare not approach this hell.

Then morning light
Brings to their sight,
A scene no pen dare tell.

Ere they returned,
Two bodies burned
To a crisp, were taken away.

They knew that three
Would never be
Found on this wretched day.

Old Captain Hill,
His body still
In this wreck's consumed by fire.

A Captain he,
Full eighty-three,
In a week he was to retire.

For many a year
The engineer
Had promptly answered the bells.

The Captain's son,
His work is done
When the water over him swells.

The sea takes toll
Of Everett Cole,
His business was in oil.

They wonder why
He had to die
When his life was eager for toil.

The sailors say,
His brother Ray

Was a wonderful engineer.

His life that day
Was snuffed away;
His body burned and sore.

Could not death wait?
Could not have fate
Passed Bruce McDonough by?

A Captain fine
At twenty-nine
He was too young to die.

Who knows the grief
From Simmon's Reef,
The anguish and the woe?

The mothers wail,
The sisters pale,
The wives are stricken low.

At fortune's side
These boys have died
Their memories are alive.

Good friends had they,
Who from that day,
Will mourn this dauntless five.

The fateful crew
Of the "Marold II"
Sailed out on New Year's Day.

Their journey done
When half begun,
Was the price too much to pay?

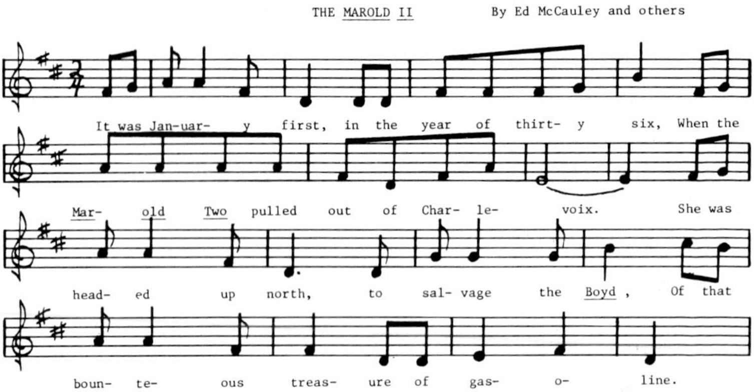

The "Marold II"
by Ed McCauley and others

It was January first, in the year of thirty-six,
When the "Marold II" pulled out of Charlevoix.
She was headed up north, to salvage the "Boyd,"
Of that bounteous treasure of gasoline.

They were five brave men, whose names I will tell,
For their glory is yet unknown.
There was Everett and Bruce, and Leon so true,
And Raymond and faithful Captain Hill.

When they arrived at the "Boyd," a cry was heard,
For there were many of her tanks to be filled.
They knew not the fate that soon was to come,
From the gasoline aboard that boat.

They were filling the bilge, when a blast occurred,
Which threw them all into eternity.
When the Coast Guard arrived, they did find her in flames,
And they knew that all were lost that day.

There were five brave men, lost their lives that day,
So take heed to my warning now:
All young women, who love your sailors true,
Don't let them sail a salvager.

The "Marold II"

In Honor of Captain Kilty

Peter Kilty, Captain of the "Pere Marquette 18," was the son of Patrick Kilty, who helped survey Beaver Island. The "P.M. 18" was built in 1902 and was 328 feet long. She left Ludington on September 8, 1910 with a cargo of twenty-nine railroad cars. That same day she had passed a government inspection. About halfway across Lake Michigan the ferry began taking water faster than it could be pumped out. The leak was never found, but Captain Kilty had the radio operator call for help and he had the lifeboats on the lee side of the ferry launched and left tied to the ferry, and the life rafts prepared. Also, some of the railroad cars were jettisoned in an effort to lighten the doomed vessel.

The "Pere Marquette 17" responded to the "P.M. 18's" distress call and arrived before the "18" sank, but was unable to approach her because of the heavy sea. Shortly after the arrival of the "P.M. 17" the "Pere Marquette 18" suddenly sank. The "P.M. 17" was able to rescue thirty-three persons, but twenty-nine went down with the ship. Survivors of the disaster praised the courage of the crew and Captain Kilty Every officer of the "Pere Marquette 18," including Captain Kilty, stayed at their posts to the last moment and went down with the ship.

In Honor of Captain Kilty
by Frank McCauley

From Ludington's fair city one dark September day,
The "Pere Marquette 18" steamed proudly on her way.
Her Captain, Peter Kilty, looked on his ship with pride,
As gracefully she glided o'er the rough tempestuous tide.

But who upon that noble ship as swiftly she sped on,
Could dream of any danger before the coming dawn;
Or before another sun would set, grim death their fate would seal,
And the dark and rolling water claim that massive frame of steel.

The Captain stands upon the deck, a seaman does draw near,
Saying "The steamer she is sinking, we cannot keep her clear."

No terror clouds his dauntless eye as he went quickly down,
Where anxious voices reached his ears, "The leak cannot be found!"

"The steamer she is sinking," is the word on every lip,
And ghastly faces filled with awe look on the doomed ship.
But Captain Kilty stands in haughty grandeur there,
Though dangers do surround him, he yields not to despair.

For in a calm and cheerful tone, he thus addressed his crew,
"Fear not, my gallant seamen, our ship is staunch and true."
"Is there a hope, or a chance for life?" a dozen lips implore,
"With God's help," he did reply, "we'll reach Wisconsin's shore."

Alas, his hopes were shattered soon, for with a mighty roar,
The seething water madly rushed and through her vitals tore,
God help those struggling creatures! Now depict it if you can,
The tragedy enacted upon Lake Michigan.

It was but shortly after the "P.M. 17,"
The wild foam rushing at her bow, arrived upon the scene.
The floating debris told the tale of that ill-fated crew,
Who to all earthly troubles had said their last adieu.

Farewell then Captain Kilty, your loss we do deplore,
Bright angels have conveyed your soul into a tranquil shore,
Where stormy waves assail you not, but pleasant scenes abound,
Where summer breezes gently blow and happiness is found.

Farewell to those poor victims who now lie in the deep,
The waves their funeral dirge do chant as they slowly o'er them creep.
While wife and orphaned children in deepest anguish mourn,
For that kind and loving father who never will return.

The Bootleggers' Funeral

The "Bootleggers' Funeral" is about the arrest of two young men — probably Frank Willmette and "Patsy Doney" Gallagher — for drinking during the prohibition era. According to the song the sheriff, after making the arrest, went back to the bar for another drink! "The Bootleggers' Funeral" was written in about 1926 by Johnny Mooney, who spent his last years in the Charlevoix Poor House.

The Bootleggers' Funeral
by Johnny Mooney

Well, Patsy and Francis one fine summer's day,
Went out hunting cattle, to the woods they did stray.
They came to a shanty at the foot of a hill,
Where of moonshine and beer they did get their fill.

There was Winters and Pat a going to town,
Taking the boys all the way up and down.
They stopped into Karnses to get a fresh cup,
Where they met with Jack Halley a lickin' them up.

They got in their car and started for town,
And Sucker Jack Halley came following down.
He nabbed Pat and Frank and did run them in,
And of course he then steered for Karnses again.

When the news went around that the youths were in jail,
And could not get back home that night from the trail,
Then two young fellows, for friendship they say,
They took a poleaxe and the lock they did break.

The day they were taken was a big day in town,
The bootleggers they all came gathering round.
There was Eddie and Jimmy and Karen so tall,
And a number of others, Old Watkins and all.

At the bootlegger's funeral there sure was a mixed crowd,
And some were quite willing to sew up the shroud.
As for the bootleggers, they had their say,
Sayin' "Now my brave boys, don't give us away."

They were taken to Charlevoix and run into jail,
Rolly Lewis's cross questions, they did not avail.
They stuck to the story, the first they did say,
'Til along came that sneak, Jack Windsors by name,
And he leaked the whole story, sure it was his own blame.

Neil Deary

I have been unable to discover who Neil Deary was. He is supposed to have composed many songs, including "The Bow of Birchie's Yoke," but apparently all of his songs have been forgotten. However, "Sissy Roddy" was Mary Roddy, daughter of Captain Roddy. She was born in 1866 and married Jack Martin, the son of James. "Kitty Strack" must have been Kitty "Strack" O'Donnell, born in 1865. There were three Mary Boyles born on Beaver Island in the 1860's, and one of them could have been the "Molly Boyle" of the song. Judging from the ages of the girls, "Neil Deary" was written in the late 1880's. Kitty Strack was quite upset about the song and supposedly demanded her name be taken out of it.

Neil Deary
by Peter "Anthony" O'Donnell

My name is Neil Deary, I'm a sailor by trade.
All parts of Americay I've already surveyed.
I came to Beaver Island, my fortune to try,
Fell in love with a young girl, and she does me pass by.

That girl I went for, she lives down the road,
At Captain Roddy's, that is her abode.
She is Roddy's daughter, as you all may know,
Her cheeks red as roses, her eyes black as sloes.

I went to this young maid, and to her did say:
"I would have you get ready, with me come away."
She looked very angry, and this she did say:
"I don't like a sailor that follows the sea,
For they say they are ragged and their clothes smell like tar,
Every cent that they make, it goes over the bar."

I'll go back to Ireland, to the old sod again,
For to better my fortune, I will cross the wild main,
And farewell Sissy Roddy, on me you went back,
Likewise Molly Boyle, and the fair Kitty Strack.

Fragments

These three songs seem to be portions of longer compositions of the Island. The first, "Mary Connolly," was by Andy Roddy, the son of Captain Roddy. He married Mary Connolly. The Connollys lived in the log home Dr. Protar later lived in. Many singing contests were once held here. "Mary Connolly" uses the tune of "Johnnie German," a song once popular on the Island. The second fragment is about Johnnie Martin of Beaver Island and is a parody of "The Fisherman's Daughter." "We Welcome You to Beaver Island" is sung to the tune of "I've Been Working on the Railroad." The last line to this song is missing, but it isn't hard to imagine what it could have been!

Mary Connolly

She is brighter than the morning star,
And fairer than the rose.
She exceeds all pretty flowers,
From the Harbor to Sloptown Road.

Johnnie Martin, So Far Away From Home

By his friends he was neglected,
And then he was rejected.
He was a poor and lonely fisherman,
So far away from home.

We Welcome You to Beaver Island

We welcome you to Beaver Island,
And wish you a long, long stay.
May the people on the Island,
Take all your dough away.

When you're broken, bended hearted,
Take the North Shore *way,*
But don't be grievous to the Island,
. . . .

SOURCES

Granny's Song. By Katherine "Kitty" Gallagher (?), translated from Gaelic by Owen "Conn" McCauley. From collection of Mrs. Helen Collar, St. James, Michigan.

Lost on Last Michigan. By Dan Malloy. From singing of Johnny W. Green, tape 60, Walton Collection, Michigan Historical Collections, Bentley Historical Library, University of Michigan, Ann Arbor, Michigan, and from other sources.

The Gallant Tommy Boyle. By Dan Malloy. Stanzas 1 and 3-7 from singing of Pat McDonough, tape 2, and the second stanza from singing of Mike O'Donnel, tape 52, Walton Collection, Folklore Archive, Department of English, Wayne State University, Detroit, Michigan.

The Beaver Island Girls. By Isaac Wright. From Mrs. Collar's Col. There is a less complete copy of this song in Box 8 of the Walton Col., Mich. Hist. Col.

The Gallant Old Mare. Author Unknown. From Mrs. Collar's Col.

The Ill-Fated "Vernon." By Frank McCauley. Box 8, Walton Col., Mich. Hist. Col.

Neil Deary. By Peter "Anthony" O'Donnell. From singing of Johnny W. Green, tape 89, Walton Col., Mich. Hist. Col.

Old Tom Sweet. By Johnnie Martin. From singing of Johnny W. Green, tape 83, Walton Col., Mich. Hist. Col.

The Fisherman Yankee Brown. By Frank McCauley. From singing of Dan Bonner, tape 60, and Pat Bonner, tape 95, Walton Col., Mich. Hist. Col.

In Honor of Captain Kilty. By Frank McCauley. Box 8, Walton Col., Mich. Hist. Col.

The U.S. "Lightship 98." By Frank McCauley. Box 8, Walton Col., Mich. Hist. Col.

The "Clifton's" Crew. By Pat Bonner. From singing of Pat Bonner, tape 23, Folklore Archive, W.S.U.

The "Clifton." Author Unknown. Box 8, Walton Col., Mich. Hist. Col.

The Seaman's Lament. By Frank McCauley. Box 8, Walton Col., Mich. Hist. Col.

The "Clifton" Tragedy. By Peter "Salty" Gallagher. Mrs. Collar's Col.

The Bootleggers Funeral. By Johnny Mooney. From singing of Ed McCauley, tape 51, Walton Col., Folklore Archive, W.S.U.

The Fate of the "Marold II." By Vaughn Ogden. Mrs. Collar's Col. This song has also been published in the Charlevoix *Courier,* Dec. 29, 1965, and in the Beaver *Beacon.*

The "Marold II." By Ed McCauley and others. From singing of Ed McCauley, tape 51, Walton Col., Folklore Archive, W.S.U.

W.P.A. Song. By Charlie "Anthony" O'Donnell. Recited by Pat Bonner, tape 95, Walton Col., Mich. Hist. Col.

Turner's Pure Bred Bull. By Clara Schmidt and Charlie "Anthony" O'Donnell. From singing of Clara Schmidt, tape 78, Walton Col., Mich. Hist. Col.

Donegal Bay. By the Allens. Contributed by Mrs. Julie Kaderabec, Ann Arbor, Michigan.

Mary Connolly. By Andy Roddy (Jr.) From singing of Johnny W. Green, tape 86, Walton Col., Mich. Hist. Col.

Johnnie Martin, So Far Away From Home. Author unknown. Recited by Pat Bonner, tape 89, Walton Col., Mich. Hist. Col.

We Welcome You to Beaver Island. Author unknown. Tape 62, Walton Col., Mich. Hist. Col.

ACKNOWLEDGMENTS

Mrs. Helen Collar provided much information about many of the songs, in addition to some of the songs themselves. Liz Cox helped prepare the music. Mrs. Julie Kaderabec contributed the song "Donegal Bay." The late Ivan Walton made this collection possible by recording Island musicians from 1932 to 1962. But for Dr. Walton's efforts, most of these songs would have been forgotten. Copies of Walton's tapes and written material are at the Michigan Historical Collections, University of Michigan, and the Folklore Archive at Wayne State University. The staff of both of these libraries were most helpful.

Beaver Island
Christian Church
NONDENOMINATIONAL

THE ISLAND'S 'LITTLE BROWN CHURCH IN THE WILDWOOD'

by Grace Hill

Volume I of *The Journal of Beaver Island History* includes a section "The Religions of Beaver Island" by Terri Hart and in it there are a few paragraphs about the infant Beaver Island Christian Church, quoted from Phil Gregg. Some of us (I and others) hoped that for Volume II of the *Journal* Phil, or someone almost equally knowledgeable, would expand on those paragraphs. My strong impression, derived from only three summers in attendance at the church, is probably entirely different from such a report but may give non-Islanders and Islanders who attend one of the other places of worship some notion of the spirit of this little building.

It's not Westminster Abbey; no sidewalk leads up to it; there are no wood carvings, sculptures, or needlepoint kneelers. So far as I know, no one famous has ever worshipped there. One might call it the "little brown church in the wildwood." The welcome mat is out to all, of whatever denomination (or of none), and no one needs to urge people, as Pitt does in his hymn, to "Come, Come, Come."

For the first years of my association with the church the "Sanctuary" measured approximately 15 x 28 feet, and not more than 30 folding chairs could be set up therein. In addition, a porch along the length of that room could accommodate perhaps another thirty people, all of whom could hear the service and some of whom could even see the preacher, providing they were in line with the open door or one of the four opened windows. Occasionally a few chairs were set up along one side of the narrow corridor leading from the entry door past the little all-purpose rooms to the sanctuary.

In the sanctuary the lectern stood beside the large fieldstone fireplace. On its mantel was a collection of items, some perhaps donated — a cross, a dried arrangement in autumn colors. A piano and small organ were probably gifts of parishioners. On walls were pictures of Christ, of two former church leaders, and a clock (the only drawback to which was that it had indicated the same time ever since I first saw it).

Now, largely in response to the burgeoning summer needs, some changes have been made, and a little of the charm has gone: the wall between the sanctuary and the porch has been removed; the clock tells time. But much is the same.

Guest preachers of varying faiths and professionalism lead summer worship. I have heard what must have been a guest's topnotch sermon (one such was the most memorable I've ever heard anywhere at any time). I've also heard what was possibly the preacher's best effort but fell far short of the mark. I've also been privileged, early or late in the summer, to hear the two year-round lay leaders. Dave Gladish, former professor of literature, now free-lance writer, woodsman, etc., gave literary analyses of a few Old Testament verses. The message of Phil Gregg, amateur historian, naturalist, marina operator, was apropos of Father's Day.

But the feature of the service which remains constant is the "joyful noise" from the congregation in its participation. A preacher's "Good Morning" is answered loudly and clearly by the congregation. And such enthusiastic singing as comes from the mouths of sixty or seventy people! The first word of the first hymn is a shock in its volume. No doubt the low ceiling of the room does something acoustically, but in any case it's an inspiring performance. Until after the last word of "God Be With You" nobody starts to pick up his belongings. All are "with it" through the last note. On some Sundays there is no Special Music. In fact, on one occasion the singing of the hymns was almost *a cappella* (and I will have to admit that my above mention of the singing would be a slight exaggeration that day). Sometimes, though, the Special Music is really special. One preacher brings his violin, which he belittles as his "squeak box," a tape player and tapes which provide his accompaniment. We have had vocal duets by amateur, but good, ladies in good voice accompanied by piano, or organ, or clarinet.

After church most of the worshippers partake of the cookies and coffee or punch served in the little entry porch; then, with cup in hand, they go on outside to the sandy ground to socialize, which leads me back to the "no sidewalk" approach. Only once did I make the mistake of wearing open-toped sandals! The sand is loose and supports scant vegetation and I do not have the knack, as some young ladies seem to have, of letting sand out of shoes as completely as it gets in. For a while repair work to the septic tank left a gravelly scar on part of the front yard. Now, some good black topsoil has been brought in, so in future years there may be more lush vegetation.

The expense of the 2½-hour car ferry trip from Charlevoix to Beaver makes one wonder how enough preachers can be attracted to give their services, without any compensation or honorarium. With their well-honed bartering sense, the Islanders have provided the incentive. For the week following each guest's services (with the exception of one Friday night per month, about which more later) the church reverts to a cottage for him and his family. The sanctuary becomes again a living room, the small rooms bedrooms, kitchen, whatever. A tiny TV, left on the mantle one Sunday morning gave evidence of how one preacher had enjoyed his leisure the night before. This free lodging at least partially offsets the travel expenses. And I understand this last year that the ferry cost for the preacher and his wife was underwritten by the church. On weekdays one sees the preacher family driving around town, or at the post office, the beach, or the ritualistic "meeting of the ferry" at the dock. If his appearance has been arranged for by a loyal member of his winter flock, now an Island summer resident, he will probably be treated to Sunday dinner, at least, or to brunch at the best restaurant on the Island. On at least two occasions guests have been persuaded to stay through two Sundays, and in one of those cases a popular young preacher had all kinds of entertainment provided — boat rides, fishing expeditions, a plane ride, and meals.

Outside on Sunday mornings there are parked eight or a dozen cars, trucks, 4-wheel drives. Some worshippers will have walked — from their boats tied up at the yacht dock or from a nearby motel. During service one hears the horn of the arriving ferry. All ages are in attendance, dressed in whatever. The guest preacher is often the only gentleman wearing a tie. One morning I noted two long skirts, one on a very young lady with her dolly, another on a beginning teenager. Another young lady in front of me "wore" also the white lines left by her swimsuit straps, now showing in the scoop neckline of her frilly blouse. Little boys are scrubbed and cleaned, albeit their all-purpose gym shoes are scuffed.

Now, about those less-than-quiet Friday nights once a

month when another of the multi-purposes of the building is served. The good smell of pancakes, fried eggs, Beaver Island maple syrup, and some variety of meat brings folks for supper, formerly for the price of a freewill donation but now for a small set price. Once during donation times, a check for $100 was found in a canister. I know of several of the year-round ladies and gentlemen who can turn out eggs and cakes which would pass inspection by the fussiest of maitre-d's. Others set tables, clear, wash dishes. Anyone available runs to one of the two grocery stores for an extra crate of eggs, or dishwashing soap, or whatever else is running low. Usually the proceeds are designated for a variety of good causes — the Boy Scouts, Girl Scouts, Civic Association (that's the local Chamber of Commerce), the Medical Center. Occasionally the church keeps the proceeds when it is necessary to replenish supplies of staples — flour, etc. — or when the well is "running dry," not figuratively, but literally. For this, of course, lots of dollars must be raised. The recent joining of the living room and porch leaves an unsightly cement strip between the two, which had been carpeted some time ago and with entirely different colors. New carpeting for this whole area would be a great improvement.

As our family waited in line on Friday evening, still outside of the building, with rain threatening, a year-round couple arrived in their truck, with the obvious intention of joining the guests. They saw the situation, eased themselves politely past the line, went to the kitchen and "set to." Their meal had to be posponed, but the guests were served promptly.

But it is Sunday morning in the Beaver Island Christian Church which comes to my mind's eye whenever I am in a place of worship anywhere, on any Sunday morning. On such occasions I think also of my one worship experience in a sanctuary which ranks at the other end of the size scale, Westminster Abbey. My companion and I sat on not very substantial, ladder-back, rush-bottom chairs, looking up at the vaulted arches, hearing the organ sounds go up and up, seemingly to infinity.

Each of these places of worship — the Beaver Island Christian Church, made possible in a few years by a few families on an island, and Westminster Abbey, built and enriched over the centuries by the British Empire — serves its own important purpose. In the vernacular, each packs a wallop.

Dr. Russell E. Palmer

Mrs. Russell E. Palmer

DR. RUSSELL E. PALMER

by Doyle C. Fitzpatrick

When Dr. Russell Palmer first came to Beaver Island in 1923, at age 30, he was "sick in the head." At least, this is what his medical friends thought. They were sincere and perhaps accurate, in relation to their own perspective. To them, comfort and the conveniences of modern living meant a great deal.

Dr. Palmer was no ordinary man. He had that certain "something" setting him apart as another self-sacrificing humanitarian. After all, he could have had a lucrative practice anywhere and lived in genuine ease.

Beaver Island in 1923 was still primitive in many ways. Its people were not primitive, but the lack of modern conveniences, common to the mainland, kept the Island population living almost as they had done during the late Mormon era and early Irish immigration. The Coast Guard had the only telephone on the Island. No electricity meant no convenient water power and no sewage treatment. Kerosene lamps were used extensively. A few people had Coleman lamps, brilliant and comparatively luxurious.

There was plenty of wood on Beaver, used for cooking as well as heating. Some islanders still use wood today and many others are now converting to wood because of the energy crisis. All homes were not primitive. All had a well and a pump; some had pumps inside as well as outside. The necessary backhouse had one seat for the anti-social. Two, three and even four seaters invariably had either a Montgomery Ward or Sears catalog serving a dual purpose. A supplement to the backhouse was a pot under every island bed.

These are some of the reasons Dr. Palmer was chided by his friends about going to Beaver. If what they said was true, then the good doctor was "sick in the head" for a commendable twenty-nine years.

Doctor Palmer was born in Canada in 1893. There he received his medical degree from the University of Toronto. Before coming to the Island he practiced medicine in Lansing. After his retirement he still considered Lansing "home" much of the time and spent several years there and in Florida where he died. In Lansing he was a member of the Ingham County Medical Association, and a member of the British Medical Association, as well.

When Dr. Palmer began his practice on Beaver Island he was not very welcome. Many old-timers resented him. Fedor Protar was still active and his services were considered satisfactory and trustworthy in spite of his advanced age and his disavowal of any medical training per se. Dr. Harry Vail followed Dr. Palmer, but due to ill health had few patients. Many Islanders were afraid of him because house calls often left him exhausted, forcing him to lie down on occasion as soon as he arrived at the home of a patient. Dr. Frank Luton, following Dr. Vail, was loved by the Islanders. He retired in 1963, due to his advanced age, and was replaced by Dr. Howard B. Haynes. Dr. Joseph Christie, who recently retired in December, 1979, followed Dr. Haynes. The present doctor is Leonard Siudara.

The practice of medicine on Beaver Island in 1923 was a completely new and different experience for Dr. Palmer. It was difficult at times. On the mainland, where practically no house calls were made, patients could get a "check-up" as often as needed in the office of the doctor. On the Island, however, nearly all patients were treated at home. This meant twenty to twenty-five house calls a day. There was no Medical Center as we know it today.

Many patients of Dr. Palmer would wait until their ailments had progressed beyond a reasonable treatment stage. Often, pregnant women would not call him until the baby was ready for delivery. There were no monthly calls to see how the pregnant patient was progressing. And too, when the woman was ready for delivery, Dr. Palmer could not leave the case. There was no hospital. There were no nurses except Mrs. Palmer. Some deliveries took two or three hours; some much longer. One delivery continued for two days.

One rather difficult delivery involved a former Beaver Island enthusiast, Mrs. Henry Allen. This birth left Dr. Palmer completely "bushed." Her husband, the late Henry Allen, always ready and willing to give comfort and aid to the doctor, asked if he would like a drink. His affirmative reply sent Henry out to the kitchen. He stayed for some time. Naturally, Dr. Palmer wondered what had happened to him. Henry finally returned holding a tray on which he was carrying the drinks. It was home brew to be sure, but only coffee!

Erratum Sheet

The following replaces paragraph two, page 120.

When Dr. Palmer began his practice on Beaver Island he was not very welcome. Many old-timers resented him. Fedor Protar was still active and his services were considered satisfactory and trustworthy in spite of his advanced age and his disavowal of any medical training per se. Dr. Harry Vail followed Dr. Palmer. Dr. Frank Luton, following Dr. Vail, was loved by the Islanders. He retired in 1963, due to his advanced age, and was replaced by Dr. Howard B. Haynes. Dr. Joseph Christie, who recently retired in December, 1979, followed Dr. Haynes. The present doctor is Leonard Siudara.

No electricity on the Island meant no X-ray or Electrocardiograph equipment. The thermometer, the stethoscope and the sphygmomanometer, for testing blood pressure, were the three basic necessities, along with urinalysis and blood testing equipment. The most important medical tools Dr. Palmer used were his hands. In Toronto he was taught that his five senses were most important in treating patients, and that his own eyes would tell him a great deal. Clinical judgment was grounded in acute observation, physical examination and deductions from these two essentials.

Common sense and humor were two assets Dr. Palmer needed on Beaver Island. He had them both. Some serious medical problems actually had hysterical aspects.

One such emergency involved a husband living in town who went to Dr. Palmer asking him to go with him immediately because his pregnant wife was in trouble. When they arrived, the poor woman was sitting on a highly decorated tall pot with her nighty around her knees. The doctor told her to get back into bed. She said she couldn't because she was stuck! As soon as he surveyed the difficulty, the doctor realized she had dropped the baby in the pot. There was nothing to do but carry the woman, the baby and the umbilical cord to the bed. The "potted kid" lived, according to the good doctor. The baby was a girl. Years later, this same girl, seeing Dr. Palmer on the street in Charlevoix, invited him to her June wedding.

An emergency requiring genuine common sense involved two women who had been delivered years before by Dr. Palmer soon after his arrival on Beaver Island. After maturity, these two women married and they both became pregnant about the same time. One of the women lived in St. James, the island's only town, and the other lived outside the village. When it was time for delivery, the one in the country was the first to send her husband for help. Dr. Palmer left at once with the nervous husband. Soon after the two arrived, there was a knock on the door. It was the other husband. Dr. Palmer suggested that the second husband go back, bundle up his wife and bring her out to have her baby there also. This he appropriately called "rural delivery."

Another brief incident upset the usual calmness of Archie LaFreniere, for many years owner of the Shamrock Bar, and now host to many tourists and gift shoppers. Dr. Palmer was called to Archie's home when Mrs. LaFreniere was about ready to deliver her first child. Labor was long; during the delay, Archie became increasingly nervous. In fact, Dr. Palmer claimed that while he was there, Archie fed his chickens eight times . . . "that the chickens never had it so good." The baby girl arrived in spite of Archie . . . is now a full-grown young lady with a little girl of her own . . . Loretta Slater, presently living with her family in Lansing.

During his early years on the Island, Dr. Palmer also cared for occupants of High Island, about four and a half miles west of Beaver. These patients were not ordinary people. They did not believe in death, nor did they believe in sickness. They belonged to a sect known as Israelites, of the House of David. High Island was occupied primarily by members of this sect. Many readers will recall their leader, "King Ben" Purnell. Ben did not live on the island himself. The occupants, all from the colony headquarters in Benton Harbor, were at one time, a thriving colony — raising large vegetable crops and exporting these, as well as considerable lumber.

In winter when it was grudgingly apparent that a member of "King" Ben's colony was seriously ill, Dr. Palmer would find it necessary to make the nine mile trip over ice by sleigh. There were also a few Indians on High Island, needing care occasionally.

One summer the caretaker of the House of David potato farm developed what turned out to be mastoiditis. He had a tremendous swelling. A colony member picked up Dr. Palmer in a motorized open boat. Upon arrival at High Island, he found, in a very small shack, four small children, in addition to the patient and his wife, and wondered how there was room for them all.

After treating the caretaker they began their four and a half mile trip back to Beaver. When they got half way across, the engine stalled. A stiff wind was coming up and it was getting dark. Worst of all, they had no oars. Luckily a fishing boat came along in time to tow them back to Beaver, or the trip might have been disastrous.

Necessity is often the mother of resourcefulness, proven by this story about a three-hundred-fifty pound fat man known as the laziest man on Beaver. He once developed a very serious carbunkle as big as a hen's egg. It was evident that something had to be done for this man. They sent for Dr. Palmer. After he arrived at the farm home and had examined the growth, the only quick solution to forestall danger was cauterization. "Does anyone here have a blow torch and soldering iron?" Dr. Palmer asked. The reply, "Yes, we have one in the barn," was a welcome answer. The iron was brought to a white heat by the torch and carefully placed near the cavity for treatment. In the morning a temperature that had been 105° was back to normal and the patient who might have died had been saved by a soldering iron and a resourceful doctor.

One more story about a grown Indian will end our Dr. Palmer stories, none of which could be related without the facts from the "patient" doctor himself, many years ago.

A full grown Indian had a very severe case of tonsilitis. After some persuasion, the Indian consented to an appointment for removal of the tonsils. It happened at this time that a medical student from Chicago was visiting the Palmers and, of course, the student was invited to go along with Dr. Palmer to the Indian's house. Mrs. Palmer, a nurse, also went along. They all got into the Studebaker and started off. As they were traveling along the road, they approached this same Indian staggering home after an all night binge, pretty well "loaded." The Indian was piled into the rumble seat and driven to his home on Louie's Point. Here he was promptly placed on a table for "kitchen surgery." When the Indian came out of his stupor he knew for the first time that he had not missed his appointment, nor his surgery, which had required no anesthetic!

This is, of course, a very brief review of some of the fine work done by Dr. Palmer on Beaver Island. The people of the Island realize they owe a great deal to him for his long and outstanding medical service. He was truly an "Island doctor" in the best tradition of the legendary "country doctor."

A doctor needs a hobby — something which is hard work you wouldn't do for a living. After 1938 Dr. Palmer became a member of the American Physician's Art Association. In 1942 and 1947 he had two oil paintings, "Three Bathers" and "Winter Landscape," in "Paregon," the Association's magazine. The Greek title means "work by the side of work." Medical men are often capable of artistic creations. They have fingers that "see" and eyes that "feel," a doubling of sensory capability.

Dr. Palmer loved Beaver Island and the people loved him. During his span of years here he witnessed many changes. His medical services, aided by the unlicensed services of Fedor Protor, met the needs of the island community well.

Response was gratitude then, which is expressed to this day. His picture, along with those of other island physicians, hangs in the Medical Center, keeping alive the memory of his practicing years.

The Beaver Island people served by Dr. Palmer were alert and talented. Nearly all Island children while he was here were taught ably by highly qualified Dominican Sisters and were graduated from the Beaver Island High School. During his service on the Island, Dr. Palmer saw four students become priests, two become doctors, one a lawyer, one a dentist and several became well trained nurses. Many island boys sailed the Great Lakes as officers and engineers.

During the Palmer era, one of the finest things said about the Islanders was this: "Anyone knowing them treated them with respect. They were proud of their Island, and had a right to be."

This sentiment then so ably expressed, is equally true today.

Note: Material for this article derives from a talk given Jan. 8, 1964, by Dr. Russell Palmer, when he addressed the Historical Society of Greater Lansing at the Michigan Historical Commission Archives, 3309, North Logan.

Dr. Palmer and friends.

THE CENTRAL MICHIGAN UNIVERSITY BIOLOGICAL STATION BEAVER ISLAND, MICHIGAN

by Dr. Matthew H. Hohn, Coordinator

Introduction

Several staff members of Central Michigan University, and Norvall C. Bovee, Vice President of Business and Finance, visited Beaver Island in the early 1930's. The uniqueness of the Island with its undisturbed habitats and isolation was perceived to be an ideal location to establish an off-campus meeting and study facility. It was at this time that 45 acres of land on Sand Bay, on the east shore of the Island, became available through the State Department of Natural Resources. This land borders Lake Michigan and provides an ideal setting for an educational and research facility. Through the efforts of Dr. Charles L. Anspach, then President of C.M.U., and Dr. Bovee, the University acquired the land in 1936.

Historical

The land was not developed until 1959 when, through the efforts of the state architect, plans were prepared and the main building was erected and named the "Beaver Island Center."

Meetings and classes of many types were held at the Center from 1959 to 1963. Many university and professional groups, including the Board of Trustees of C.M.U., held meetings and workshops. Several off-campus classes, including Ecology, Ornithology, Art, and Recreation, along with workshops in bus driver education and secondary education, were held at the Center during the summers. As a result of these classes, it became evident that the habitats on Beaver Island and their close proximity would provide excellent opportunities for field studies. Application was made to the National Science Foundation (N.S.F.) and a grant was received to support an institute for secondary school teachers of biology to further their studies and become more efficient in field biology.

The first classes of this type, eight weeks in length, were held during the summer of 1964. A general purpose laboratory building was constructed in 1964 for use in class lectures and observations through the microscopes. Thirty-two participants, twenty-eight men and four women, attended this institute and the one offered in 1965. When a three-year sequential institute was approved for 1966-1968 with forty-eight participants, it became evident that the sessions were successful and that additional space would be necessary for laboratory use. Two laboratories were added in 1966 so that two classes could be held simultaneously. The third laboratory was used for library, storage, and reference specimen collections.

In 1966, the board of trustees of C.M.U. officially designated the center as the Central Michigan University Biological Station and appointed Dr. Matthew H. Hohn the first director.

Students in classes not supported by the N.S.F. institutes were encouraged to attend the station, and by 1969, when the institutes were discontinued, sufficient students attended the station to permit continuation of its operation. More and more students pursuing the degree of Master of Science were attending the station. For this reason a fourth laboratory was added to be used primarily by graduate students and staff.

During the last few years of the institutes supported by National Science Foundation grants, married students requiring living space, put great pressure on the resources of the State Forest Campground adjacent to the C.M.U. Biological Station. It was then that a C.M.U. campground was established just west of the main station buildings. Electricity, shower and toilet facilities were provided and platform tents were erected for student and staff families.

The greatest use of the station started in 1971 when the May mini-session was established, along with credit and non-credit sessions for high school students. Several groups other than biology also started to use the station during the month of August. These included community college teachers through a N.S.F. grant, Lloyd Shaw Dance Foundation, and more recently, Athletic Training.

The latest additions to the station were in 1979 when the kitchen was enlarged and two housing units were added for faculty and staff.

Special Studies

The undisturbed habitats of cedar swamps, sand dunes, beech-birch-maple climax forests, beaver dams, and inland lakes provide suitable environments for class use and special studies. The range in water conditions from acid to alkaline, and the presence of bogs on the lakes provide a wide distribution of ecological situations for aquatic biology and water chemistry. State-owned Garden and High Islands also are excellent undisturbed areas for ecological investigations and their close proximity permit extensive use.

All of the students in the N.S.F. institutes were required to do a field research study. These, along with similar studies by other students, have resulted in approximately 200 research papers that are on file at the station. Studies dealing with entomology produced an excellent collection of the insects of Beaver Island with many of the papers covering their behavior and distribution. The plant studies include orchids, woody plants, bog plants, and aquatic plants from the lakes and beaver dams. A reference herbarium collection of plants is available at the station for use by interested persons. An extensive fungus collection has also been prepared by the mycology classes. An up-to-date list of birds for Beaver Island is also available. Research programs dealing with fish in Fox and Barney's Lakes and in Lake Geneserath, along with small mouth bass population analyses from Indian Harbor on Garden Island are also obtainable.

Approximately twelve theses for the Master of Science degree have been produced. These definitive studies include topics dealing with algae, smallmouth bass, insect larvae, fish populations, aquatic plants, grasses, and garter snakes.

Several of the biology staff at C.M.U. are carrying on research on a continuing basis. These include studies in geology and rock identification by Richard Dietrich; algae by Daniel Wujek; shore bird populations of Beaver and surrounding islands by Michael Hamas; and the flora of Beaver Island by Matthew Hohn. All of the publications are on file at the station for use by interested persons. The *Flora of Beaver Island: Part I The Bogs* may be purchased at the book store at the station or from local merchants and the Historical Society. *Part II The Lake Michigan Beaches and Sand Dunes* will be available during the summer of 1980.

There is a wealth of unpublished data concerning Beaver and the surrounding islands that are available for general use by residents and visitors. Interested persons may use the library and all reference collections of birds, plants, insects, fish, and fungi. The station is available for non-profit organizations and educational groups when classes are not in session.

The highlight of the season for Island residents occurs in October of each year, when a closing-the-station party is held. All Island residents are invited; everyone who can possibly make it, attends. And why not? The huge open-hearth fireplace, glowing with hospitality, is a welcome sight as one enters. Quickly, the large common room fills up with happy Islanders who always look forward to this annual occasion.

The facilities of the station, for one evening at least, are completely turned over to the Islanders. Everyone is comfortably seated for dinner (and what a dinner!). Music and dancing usually follow, lots of fun and chatter go on, with some folks already referring to "next year's party!" You can be sure the C.M.U. Station is an integral part of Island life.

THE BROTHERS' PLACE

by Brother Edward Fallon

For over fifty years the Christian Brothers have been visiting the Island during the summer months. A generous donation by J. P. Maloney instituted this tradition. The planning, construction and maintaining of Brother Domnan Lodge (the Brothers' Place on Beaver Island) was made possible by the generosity of many God-fearing people, Islanders and Outlanders alike. This short history touches a few of the memories that make up the Christian Brothers' tradition on Beaver Island.

Beaver Island

by Agnes Maloney Smith Payne
(Daughter of J. P. Maloney)

Serenity and peace.
Pines whispering in the night,
A moon that dips its silvery face
Across the bay in shimmering light.

Warmth and friendliness.
A sharing of each sorrow,
Island of enchantment.
You help me meet tomorrow.

For over a half a century the Christian Brothers (Brothers of the Christian Schools) have made their home on Beaver Island. During these years their presence has become part of the lore of Beaver Island. If it were not for the leadership and vision of several people the Brothers would never have set foot on the island.

Chicago was the place where contact was made which eventually led to the Brothers' presence on Beaver Island. The old St. Patrick's Church located on DesPlaines and Adams had a choir director by the name of Dr. J. Lewis Browne. Earlier he had taught in the East. He was on the staff of the New York Protectory, an institution dealing with the young who were wards of the State of New York. The Brothers managed that facility for many years. The Director of the old St. Patrick Academy was Brother Herbert Lewis. Dr. Browne and Brother Herbert were conversing one day and Brother Herbert mentioned to Dr. Browne the interest the Brothers had in locating a site for a new facility for young men interested in becoming Brothers. Dr. Browne said if he heard of any possibilities he would contact Brother Herbert.

The following Sunday Dr. Browne was the guest of Mrs. Tracy Maloney, a choir member in St. Pat's Parish. Four were there for dinner, Mr. and Mrs. Tracy Maloney, a J. P. Maloney and Dr. Browne. In the course of the conversation J. P. Maloney spoke of a religious group having refused the land he offered on Beaver Island. Mr. Maloney stated he was willing to give it to a religious group. Not wasting any time, Dr. Browne shared the news with Brother Herbert who, in turn, contacted Mr. Maloney. Both Dr. Browne and Brother went to Hub and visited with J. P. Maloney and Mr. Malloy, who were natives of Beaver Island.

J. P. had controlled Booths Dock when he lived on the island; his sister taught school. Curiously, the Christian Brothers' association with schools and J. P.'s sister being a teacher, led to an interesting and warm association. Mr. J. P. Maloney stated he was most willing to give the property to the Brothers. The time was late May of 1928.

Brother Herbert contacted his superior, Brother Leopold Julian, who was at the Brothers Provincial or District headquarters in Glencoe, Missouri. Through the means of a long distance phone call, Brother Herbert was given the go-ahead to travel to the island and inspect the property. Brother Justin Clarence and Brother Herbert left Chicago for Beaver Island. Since the school term was finished the Brothers had the leisure to stay at the then Beaver Hotel and gather data on the island and inspect the property.

Family ties are strong on Beaver Island. The Brothers had no difficulty inspecting the property because J. P. had contacted his brother-in-law, Joe Burke, and Tracy Maloney contacted her brother, Lawrence.

Joe Burke, Brothers Herbert and Justin were given a complete tour of the property which comprised forty acres. Brother Herbert chose the northern twenty acres. His reasoning seemed to be the natural cut off from the Kings Highway was separated by a natural barrier of trees.

Mr. & Mrs. J. P. Maloney's generosity was finalized through a deed dated in May, 1928. The generous gift was

Brother Francis F.S.C.

Brother Herbert F.S.C.

given "for the use and benefit of the Christian Brothers connected with the school (St. Patrick Academy) conducted by the Order in the City of Chicago." In short, Elizabeth and J. P. Maloney donated the property for whatever use the Brothers desired. Later, another twenty acres was donated by Elena and Mary O'Leary. This deed is dated October 10, 1957.

Interest in Beaver Island on the part of the Brothers grew. The St. Patrick Brothers Community and the then St. George Community of Evanston, Illinois traveled to the island. They boarded the ship "James Sanford" which used to ply the waters between Charlevoix and St. James.

Brother J. Elzear, Herbert Lewis and Liguori Francis discussed the possible use of Beaver Island with their superiors, Brother Leopold Julian and Abban Philip. This discussion lead to the approved plan of building a combined retreat house and summer home for the Brothers.

How the Brothers could build a facility on the island was not an issue. The bigger task was soliciting support in Chicago for the project.

When the Brothers, especially Brother Herbert Lewis, shared the dream with the St. Pat's Alumni, a Mules Devine and Steve Dolleyand came forward and took the leadership, calling themselves the "Ol' Timers."

"Ol' Timers" was appropriate for they wanted to have a testimonial dinner for one of the "old" teachers, a Brother Domnan. Brother Domnan was Canadian by birth. He was up in years. The archives state, "Brother Domnan means as much to the youngest graduate as the oldest." Instead of a separate testimonial to Brother Domnan the Alumni closed ranks and had their Annual Alumni Banquet honoring Brother Domnan.

When the "Ol' Timers" and the alumni heard Brother Herbert's dream of a house on Beaver Island for the Brothers, the group reacted favorably to the plan. They suggested a testimonial in the form of a summer house be made in the name of Brother Domnan for use by all the Christian Brothers.

A definite plan was announced in November of 1928 at an annual St. Pat's Alumni Banquet. This banquet was held in Chicago at the Palmer House. The plan called for a home "in the neighborhood" of $5,000. It was to be known as the "Brother Domnan Testimonial Lodge." When Brother Domnan died on April 2, 1929 the name was officially changed to the Brother Domnan Memorial Foundation.

Names like General John J. Garrity, John H. Penny, Philip J. Kurtzenborn and Peter J. Crowley were, as the early history read, "appointed" to see that the testimonial would become a reality. This group was assisted by William T. Rigney, John J. Gouhy, Eugene P. Williams and William T. Fortune.

Another group was very generous with their services and money. Monsignor Edward J. Fox, Robert M. Sweitzer, John S. Keogh, Morgan A. Collins, Daniel J. O'Connor, John P. Curran, James T. Prendergast, Thomas J. Hickey, James J. O'Grady, Michael S. Kerwin, James S. Palmer, Dr. Thomas P. Foley and Joseph A. McMahon.

The "Ol' Timers" had a younger counterpart known as the "Iron Men." They were particularly active in soliciting and contributing donations. This organization visited the island during the Thirty Day Retreat the Brothers had in 1929.

As the Testimonial Fund Drive evolved under the leadership of Brother Liguori Francis, so did the original building plans. A smaller structure was planned, discussed and just about agreed upon when Brother Herbert observed a larger structure could be used for Thirty Day Retreats for the Brothers. Brother Herbert's observation was implemented into the general revised plan. The plan was upgraded to house fifty Brothers.

The "Ol Timers," "Iron Men," many alumni, friends, Brothers Herbert Lewis, Liguori Francis and Honorius Patrick all successfully solicited funds. Boxing Shows, penny raffles, and weekly collection within the Brothers High School at St. Pat's in Chicago assisted in making the Brother Domnan Memorial Lodge a reality.

Construction Of The Buildings

The final building plan was an adaptation drawn up in 1928 by Brother James Francis and a Charlie Walsh or Wallace of Joliet. The original design was simple. All lines were in the form of a rectangle or square. Brother J. Elzear, a Brother stationed at St. George High School in Evanston, Illinois, suggested the two wings should slant a little instead of having a straight architectural line from one end of the building to the next. This suggestion was implemented. The Chapel and Common Room or living room section angled toward Kings Highway.

Team work was obvious in the planning stages of the memorial. A Joe King, Assistant Principal of Harper Junior High School in Chicago, drew a pencil sketch of the floor plan with its general dimensions. This floor plan was used by Joe Burke, the head-carpenter and general supervisor of the construction. Joe was a brother-in-law to Mr. J. P. Maloney, donor of the property. As Joe oversaw construction of the lodge many changes were made. The quality of construction was excellent. Measurements were accurate, materials were good and the end product today is a tribute to that original construction crew.

July of '28 found the concrete foundation being poured; the following year the present two story structure became a reality. As soon as the main building was finished the Brothers held their first Thirty Day Retreat in 1929. While the Brothers were on Retreat they helped paint the new structure. Joe Burke was no sooner finished when he began construction on the dining room. The Brothers worked under the direction of Joe Burke when the garage was being built.

The plumbing was planned by Mr. Powers, who had a son graduate from St. Pat's in 1929. His interest and concern were evident since he made four different trips to the Island. On one of his trips he brought his entire family. On another,

Brother Herbert and P. J. Crowley accompanied him. The on-site plumber was "Barney" Chambers of Charlevoix assisted by James Keating, a 1925 graduate of St. Pat's. Jack of all trades, Joe Burke, drove the point for the well thirty-six feet deep before he hit water. There were fifteen feet of standing water in the pipe. The water system changed three times over the years. Originally Brother John Nicholas rigged up an old Ford engine for the water system. As the years unfolded the engines or pumps were upgraded. Sewage was handled by a septic system consisting of two septic tanks located at either end of the building.

The lumber used in the Lodge came from the East Jordan Lumber Company of Petoskey, Michigan. Most of it was shipped on the "Rambler," a freighter. A Mr. McManus of Charlevoix was the man responsible for all the mainland arrangements for the lumber and building supplies. The old Parmelee Company of Charlevoix did furnish some of the building materials.

Oil lamps were the main source of light during needed times. The Delco and Kohler manufacturers would not guarantee their power supplies so having a "lamplighter" or several of the Brothers tending the lamps became an annual event until the '40's when Brother Honorim Edward wired the lodge for electricity using hand-me-down fixtures from the Brothers' schools in Chicago.

People Programs

Historically, the presence of the Christian Brothers on Beaver Island spans fifty years. During that half a century many social, spiritual and working relationships were established. Given the spirit of the thirties, the Brothers thoroughly enjoyed the Island Recreational outings and Spiritual Retreats which were annual events. Friendships grew strong.

Christian Brothers from the Midwest felt at home on the Island. The LaFrenieres, McDonoughs, Gallaghers, Boyles, Maloneys, Culls, McCauleys, Coles, Gillespies, Greens, Martins, Allens, O'Donnells, Bonners, Burkes, Gatliffs, Wojans, Malloys, McCanns, Nackermans, Annands and later the Wagners, Townsends, Greggs, Christies, Welkes, Fitzpatricks, Beckers, McGlocklins, and the Burris' were some of the litany of names that made Beaver Island special to the Brothers. The old annual four o'clock in the morning passage from

Beaver Island to Charlevoix brought many of the Islanders out to bid the Brothers good-bye at the end of their spiritual renewal programs.

It was rewarding to see the way the Islanders and the Brothers increased their fellowship. As the Brothers grew in numbers in this country, the use of the Island facility became more popular in the Order and more widely known. After the Second World War the twenty and thirty day Retreats became one hundred days. Brothers from across the country and Canada came to the Island to renew their ideals and relax. To this day a Christian Brother who made a Retreat on the Island always speaks well of the Islanders; how warm, hospitable, friendly and understanding they were and are. When I was in Rome in 1971 several Brothers from Canada reminisced until early morning about their experiences on unusual Beaver Island.

Frequently the Brothers in charge demanded much from the Brothers when they spent time on the Island. These demands were lightened by the friendship and social life offered by the Islanders.

Brothers Cornelius, Janes Francis, Herbert Lewis, Luke, George, Richard, Edward, Gabriel, Emmett, John, Kevin, Michael, Aloysius, Frederick, Joseph, Cyprian, and Fabian were among the regular summer dwellers for the various time spans of history.

Changes in the Catholic Church and Society gave rise to influences within the Brothers. The successful programs of renewal on the Island gave birth to a new facility in Sangre de Cristo, New Mexico. This became a year long program.

During the mid-sixties the Brothers' use of their facility on the Island was limited. Administratively their mid-west part of the country was undergoing change. The original St. Louis District or Province was divided into three parts, namely, upper midwest (Minneapolis, Wisconsin and the

Brother Domnan.

Plains States) and the Chicago area (Michigan, Indiana and Illinois) and the South, St. Louis (Missouri, Tennessee, Ohio, Oklahoma, Kentucky and Kansas). This division eventually put the Island retreat under the jurisdiction of the Brothers from Chicago. In great part it re-emphasized the origins of this serene haven and how the Brothers came to the Island.

Today the Brothers, their relatives and friends enjoy the Island annually. The spirit of the "Ol' Timers" lives on in Brother Domnan Lodge — "The Brothers' Place."

THE BUILDING OF THE HEAD LIGHT HOUSE

by Helen Collar

This article deals with one of the earliest and most amusing real estate miscalculations ever to arise on Beaver Island. The successive occupations of the Island by differing groups and the manner of their dispersals have played a part in many a land-holding mix-up. But the tale which follows involves the U.S. government and therefore affords us the chance to speculate on how little the passage of time has altered the ways of the bureaucrats in Washington, or the deviousness of land speculators where they see a chance to gain a buck.

In the middle of the nineteenth century, navigation on Lake Michigan was steadily increasing. In Washington it was decided that a lighthouse was needed on the Beavers to aid the growing number of ships which were sailing between Chicago and the Straits of Mackinac. The result was the lighthouse at the head of Beaver Island, and the story of its building is so full of bungling by bureaucrats and chicanery on the part of speculators that it bears telling.

It began with the laying aside by President Fillmore of a tract of land for the proposed lighthouse. This took place on November 21, 1850, and the area designated was the 158.2 acres that comprise Cheyenne Point. On the following January 7th a diagram of the tract was sent to Washington and approved. On March 12th a contract was drawn up at Mackinac Island between Charles E. Avery, Collector of the District of Michilimackinac, and John McReynolds of Detroit, McReynolds agreeing to build a tower, a light, and a dwelling for the sum of $4,480.00. The contract included detailed specifications for the structure to be built, and stipulated that it be completed by October 1, 1851. The building was to be erected on a spot to be designated by Charles Avery. The agreement was signed by the contracting parties and witnessed by James T. Birchard.

The lighthouse was finished early in October as specified in the contract, and McReynolds was paid October 21st. At that time there was no question raised as to the location of the building, and Avery and McReynolds seem to have been blissfully unaware that they had made an astonishing mistake; for land records show that the lighthouse they had just finished stood more than two miles east of the area approved by Washington. The exact site had been the responsibility of Avery. Had he picked the wrong place, or had McReynolds misunderstood his orders? Whatever the reason, the land on Cheyenne Point had not been used, and the Head Light was erected on Lot 3, Section 20, Township 37 north, Range 10 east.

This should not have been too disastrous a mistake for, according to the records in Washington, Lot 3 had not been sold, and the government owned the land it had used even if it was not that originally assigned. Although Avery and

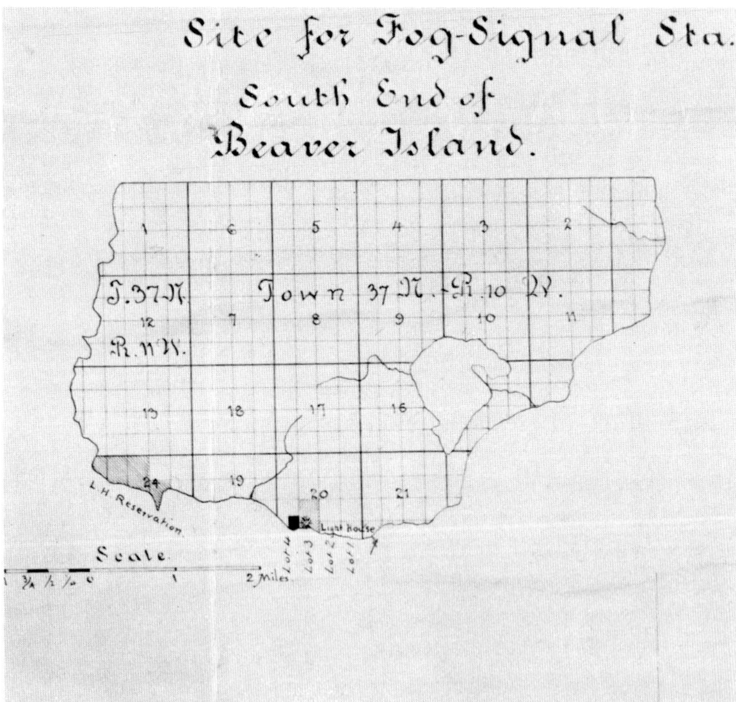

McReynolds were unaware of the mistake they had made, there appears to have been one man who realized that Lot 3 was not Cheyenne Point. We have seen that the contract for the building was witnessed by James T. Birchard, and the Birchard family are the heroes, or the villains, of this story.

The land records of Beaver Island contain the names of James T., of Matthew, and of Lorenzo Y. B. Burchard. The census records show that Lorenzo and his wife Amanda were living on Mackinac Island in 1850. James and Lorenzo appear to have been brothers, or cousins, and Matthew was probably the son of Lorenzo. In the material on the lighthouse the name is spelled Birchard. The keepers of early records were easy-going about accuracy in the spelling of names. This results in regrettable confusion; but Lorenzo Y. B. Birchard and Lorenzo Y. B. Burchard must be the same man. Because of this cavalier regard for spelling, one can not help wondering if the Burchards were not also the Bouchards, a family that was on Mackinac early enough to have been recorded in the 1830 census under the name of Eloi Bouchard. Although it was never their home, the Burchards became extensive real estate dealers in Beaver Island land, their first transaction occurring in November 1851, and the last in 1892 when the estate of Matthew Burchard was liquidated.

The first transaction, that of November 1851, shows Lorenzo Y. B. Burchard selling Lot 3, Section 20 to James T. Burchard; but there is no record of how or when Lorenzo acquired the land he was selling. This purchase took place one month after the lighthouse had been completed. The next summer, July 1852, James sold Lot 3 back to Lorenzo, and from then on James, the one who had witnessed the contract and who, presumably, was the first to realize that the lighthouse was in the wrong place, disappears from the picture. From that time on the affair was solely in the hands of the man with the fascinating name, Lorenzo Y. B. Burchard. These sales between the Burchards are a puzzle when we realize that they took place at a time when, according to the Tract Books of the Land Office in Washington, Lot 3 was still owned by the United States government.

In the fall, after his repurchase in July, Lorenzo apparently told the authorities on Mackinac Island that he, not the United States, owned the land on which they had erected their fine new building. Evidently he was believed, for on October 6th Burchard and his wife Amanda, for the sum of $150 made over to the United States 10 acres of Lot 3, reserving for themselves fishing rights on the beach and off shore. This was a tidy profit, for the 10 acres sold were approximately 1/6th of the total area of the lot, a 58.85 acre tract, the whole of which at the going price of $1.25 an acre was worth only $73.56. As we shall see later, there is some question whether Burchard had at this point paid anything for the land, and it also appears from later developments that he received no money at this time.

Lorenzo, whose later Beaver Island land dealings show him to have had the heart of a speculator, must have felt pleased with himself on that October day when he made his sale in Mackinac; but he was to find that it was easier to sell land than to collect the money, at least from the United States government. Thirteen days after the sale, on October 19th, the Superintendent of Lighthouses at Mackinac received a letter from the Secretary of the Lighthouse Board in Washington saying:

> In relation to the tract on which the Beaver Island Lighthouse is built, . . . when the jurisdiction of these premises for the United States is approved by the Attorney General, you will authorize the payment of $150.

This letter was only the first of a flourishing correspndence involving Washington, Detroit, and Mackinac Island. For the next four years letters went back and forth, all concerned with payment for the land on which already stood a lighthouse that blinked its signal nightly, guiding the ships and sailors of Lake Michigan. The Treasury Department refused to issue payment until they saw a valid title. Finally on January 2, 1855, after too much correspondence for us to record here, the office of the Attorney General of the United States sent a letter to the Secretary of the Treasury enclosing an abstract of the land in question. This abstract had been validated by the United States District Attorney's office in Detroit and was dated November 14, 1854. Sixteen days later the same office sent another letter validating Detroit's validation.

On March 31st of the same year a frustrated Lorenzo Y. B. Burchard appeared in person at the office of the Lighthouse Inspector in Detroit, and asked where he should apply for his money. He was evidently told he must give proof of ownership, because over a year later, on January 3, 1856, the Inspector wrote:

> The person claiming the land on Beaver Island promised to forward proof of his title, but I have not heard from him.

It seems that Lorenzo did not send proof because he had none. Between his visit to the Inspector in March 1855 and

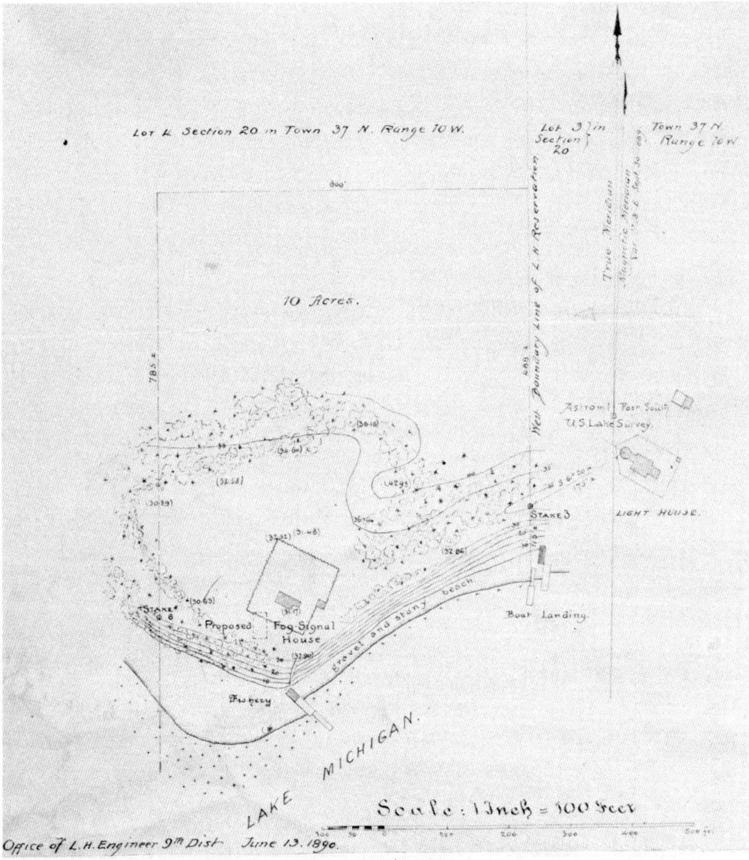
Lot 4 Section 20 in Town 37 N. Range 10 W.
Lot 3 in Section 20
Town 37 N Range 10 W
10 Acres.
West Boundary Line of L.H. Reservation
True Meridian
Magnetic Meridian
Astrom'l Post South U.S. Lake Survey
Stake 3
LIGHT HOUSE.
Boat Landing.
Stake 6
Proposed Fog Signal House
gravel and stony beach
Fishery
LAKE MICHIGAN.
Scale: 1 Inch = 100 Feet
Office of L.H. Engineer 9th Dist June 13. 1890.

August 1856 he must have decided that if he was to get his money, he must buy that for which he was trying to collect payment. On August 27, 1856 came the first entry of this land in Washington. The Tract Books of the Land Office, Department of the Interior have this record:

> Lorenzo Y. B. Burchard paid $73.56 for Lot 3, Section 20, Township 37 north, Range 10 east.

This leaves a puzzling question, how did those supposed transactions of 1851 and 1852 get into the County Records when they do not appear in the United States Land Office books? It is on this uncertain note that the saga of Lorenzo Y. B. Burchard and the Beaver Island Head Lighthouse comes to an end.

There is a footnote to be added. One hundred and three years after the setting aside of Cheyenne Point the mistake was rectified that had been made in 1851. Like those of the gods, the mills of bureaucracy grind slowly; on August 19, 1954 President Eisenhower rescinded President Fillmore's order of December 21, 1850.

Freidrich Parrot
(F. Protar)
c.1870

PROTAR CHRONOLOGY

Born: March 6, 1838, Dorpat (Tartu)
Confirmed: July 24, 1855, Lutheran (17)
Studied: March 1857-1863 (?), Dresden (19-25)
Married: before 1869: Anna Strutz (31)
America: before December 1874 (36)
Rock Island, Illinois: 1882-1893 (44-55)
Beaver Island: 1893-1925 (55-87)
Died: March 3, 1925, Beaver Island (87)

PROTAR REVISITED

by Antje Price

Since publication of the first chapter on "Dr." Protar in Volume I of the *Journal of Beaver Island History* (1976), much new material about him has come to light, mainly through contact with his family in Europe. This chapter is being written to update, correct, and supplement the original one with data from these new sources. It will not repeat the material given in the first volume, nor will it deal with information being gathered now on the Island, an ongoing project to be reported in a future issue of the *Journal*.

New Sources

In the spring of 1977 a Mrs. Barbara Robinson of Munich, Germany, wrote to the Beaver Island Historical Society to inquire about material regarding her relative, "Protar." She had heard about the possible existence of this through a chance conversation with an American exchange student who had read Ludwick's *Russian Doctor in Paradise* when studying Great Lakes history in high school in Chicago. A lively correspondence ensued with the writer of this chapter, as well as two visits (1978, 1980) to Munich for exchange of information with Mrs. Robinson, great-granddaughter of Protar's uncle Wilhelm, and Irene Treboux, granddaughter of his brother Otto. All this communication was in German. Along with a great deal of other valuable material, Mrs. Robinson provided a number of photographs, including some from Beaver Island that Protar had sent to the family in the early 1920's, and those included in this chapter. With her help, contact was made with Tartu (formerly Dorpat) University in the present Soviet Republic of Estonia, which generously sent a wealth of material about Protar's ancestors and immediate family, most of which, however, is still awaiting translation (from Estonian and Russian). American library sources have also yielded some new material, as did the Charlevoix County Probate Court records.

Ancestors

The information gathered earlier from various sources about Protar's illustrious forebears (see Vol. I, pages 54-55) was confirmed by the family. In addition, it was learned that his other grandfather, Johann Wilhelm Krause (1757-1828) was an equally ingenious and prominent academician who had also emigrated to Livonia (Estonia) at the turn of the nineteenth century. Originally from Upper Silesia, he had enlisted in the army of a small German principality (Anhalt-Zerbst) when he had to end his education due to financial reverses, and was sent with his regiment to fight as a British mercenary ("Hessian") in the American War of Independence. He reportedly resigned because he "did not want to fight against freedom." While in America and during later travels in Europe he studied agriculture and taught himself architecture. From 1803-10 he designed and erected the buildings of the new Dorpat University, many of which are in use and much admired to this day. He then became professor of architecture at the university, which was under the leadership of Georg Friedrich Parrot, its first chancellor and Protar's other grandfather.

Alluded to but inadvertently omitted from the information about Protar's father, Dr. J. J. Friedrich W. von Parrot, in Volume I of the *Journal,* was the fact that he was the first man to climb Mt. Ararat, which he did on a scientific expedition in 1829. He was also the first to climb Monte Rosa (1816) in the Italian Alps and Mt. Maladetta in the Pyrenees in northern Spain (1821). The purpose of these expeditions, as well as one to the Crimea and Caucasus, for which he interrupted his medical studies in 1811, was to measure land elevations and sea levels, a new science at that time. This was done with a nivellier instrument, a kind of barometer which Parrot had perfected for himself in 1816, and which went into Armenian legend as "the very clever spirit in the box," consulted by Parrot from time to time as he climbed. He was also remembered in the Ararat region as a wizard, for whom "silver rubles grew like mushrooms under the bed," although he had meager funds, having financed the expedition himself and not being reimbursed by the Russian state until his successful return, when he was given the Order of Ste. Anne, including the "von" before his name. The measurements in Italy and Spain were the first ones done by one man alone. On the latter he determined the relative levels of the Atlantic and Mediterranean seas. It was for the remeasurement and correction in 1829 of his earlier findings (1811) about the levels of the Black and Caspian seas, that is, his admission in print that he might have been wrong, virtually unheard of

J. J. Freidrich W. Von Parrot
(1791-1841)
Protar's Father

Emilie Parrot
(1798-1866)
Protar's Mother

from a scientist in that day, that Alexander von Humboldt praised him. In all the travels, which were done on foot and often alone, he studied the local geology, botany and ethnography. Like his father, he had a number of inventions to his credit, as well as medical discoveries, detailed in a wealth of published papers.

Livonia 1838-1857

The dates, places and names cited in regard to Protar's birth, confirmation and marriage in Volume I of the *Journal* are erroneous and should be disregarded. They are based on entries in a Bible that had belonged to Protar and is now in private hands on Beaver Island. That information refers to Protar's oldest brother Moritz (1831-1882). It is assumed that the Bible was sent to Protar after Moritz's death, possibly by his widow Helene, although this second marriage was not recorded in it.

The man who called himself "F. Protar" in America was born Friedrich Pirs Parrot on March 6, 1838, the third and youngest son of Dr. Friedrich W. von Parrot (1791-1841) and Emilie Krause (1798-1866). He was not yet three years old when his father died after three years of illness. His mother had to bring up her sons alone on a widow's pension and often had money worries, there being no "large Russian estate" as legend would have it. They continued to live in Dorpat (Tartu), although the Parrot houses at 6/8 Riia Street had been sold a few months after his father's death. His famous grandfather, Georg Friedrich Parrot, was apparently not living in Dorpat at this time; he may have remained in Petersburg after retiring from the Academy of Sciences there in 1840, and later was known to be in Helsinki, Finland. At any rate, his influence on the fatherless boys does not appear to have been very strong.

Exerpts from Emilie Parrot's diary (1844-66) provide some valuable insights about the school years and early manhood of her youngest son Fritz, as he was known in the family and as he will be called in this part of the discussion. He caused her much worry, beginning with school problems in 1864 when he was eight: "He has my flightiness" she wrote. Three years later he had to leave the school because "his laziness is holding back the other students," but he did no better in his new placement. His brother Otto, a year-and-a-half older, was also not working up to his capacity. After much consultation, the boys were admitted to the Dorpat gymnasium (high school), having passed the entrance examination, Fritz in the 7th grade and Otto in the 8th. But school problems are mentioned again the next year: "both (boys) don't want to study, (they) dislike their teachers." However, they were promoted to the next grade after two years at the school. In early 1853 Fritz came down with the "cold fever" (?)

that seemed to plague him every five months, and missed some school. Neither of the boys was promoted that year. In 1854 they were taken out of the gymnasium and sent to a boarding school in Fellin (Viljandi) about 45 miles due east of Dorpat, apparently their first time away from home and their mother. This move serves to explain the conversation reported by Patrick Bonner (on tape), where Protar is said to have stated that he was sixteen years old when he left home.

"God, you left young!" Pat said, "Why did you leave so young?"

"Oh, didn't like the rules. Too overbearing!" Protar re-

Otto Parrot
Protar's brother

plied, in Pat's imitation of his voice. It had always been assumed that he meant Russia and the Russians by this remark, but it now seems clear that he was referring to leaving home for boarding school because of problems with the strict and demanding regimen so typical of the gymnasium in the German tradition.

At any rate, the boys settled down at Fellin and made the most of their abilities, finishing the school year with good reports for the first time in years. No further school problems are mentioned, despite Fritz's serious illness (dysentery, typhoid fever) in the winter of 1855-6 which kept him out of school for three months. Furthermore, in spite of failing to be promoted twice in the lower grades, both boys finished high school at the appropriate age. It is felt that the difficulties in school were probably a reaction by two bright and lively boys against dull and rigid academic regimens that offered no latitude and encouragement for their development, which was not the case at the more humane school in Fellin. It is understandable that these two brothers were very close, having gone through so much together, as well as being near in age.

During summer vacations (June-August) the small family (mother and two sons) traveled to the shore of the Baltic or to visit relatives, including the boys' half-sister Berta, their older brother Moritz in Reval (Talinn), their uncle Wilhelm in Burtneck (about 75 miles SW of Dorpat), and in 1851 their 84-year-old grandfather G.F. Parrot in Helsingfors (Helsinki) Finland, the year before his death. At sixteen Fritz traveled alone (but apparently with a friend) for the first time, to Burtneck. The next summer the boys were confirmed in the Lutheran faith, after a month of instruction (1855).

Europe 1857-1874

In early 1857, his basic schooling completed, Emilie sent Fritz to Burtneck to obtain his uncle's blessing before he went out into the world, Wilhelm Parrot being not only a Lutheran minister but also "my husband's only and highly esteemed brother." At this time he may also have received his father's little prayer-book, since his entries in it date from that year. Otto was already studying at Dorpat University. In mid-March he set out on a week's journey to Dresden in Saxony, to begin a six-year course of study in civil engineering at the Polytechnic Institute.

There is a seven-year gap in Emilie's diary entries from

this date. It is surmised from his notes that Fritz traveled considerably during this time, but not back to Dorpat as had been supposed earlier. He apparently completed his intended course of study in due time and then turned full-time to the stage. Whether he was involved in the dramatic arts while a student at the Polytechnicum, that is, exactly when his acting career began, is not known, but it is assumed that this may have been quite early, since Dresden was noted for its theater and he would have needed an apprenticeship before he could hope to earn his livelihood on the stage, which he apparently did as soon as he left the technical school. In fact, he may have chosen to study in Dresden for that reason, with the civil engineering, which was respectable but which he never used, only an excuse to get him there. In view of his family tradition and standing in Livonia, it would have been impossible for him to follow his vocation as an actor in Dorpat. He did not return professionally to the Baltic area until he was appointed to the respectable position of "Artistic Director" in Riga, after his mother's death and significantly under his own name; as an actor he used the stage name "Piers."

When Emilie's diary entries resume in 1864, Fritz is no longer in Dresden but on the road, in Würzburg (Bavaria), visiting relatives in Stuttgart, and then Königsberg (East Prussia), Carl Bernhardi's home town and where he was a university student at that time. Fritz returned home at the end of May "after 7 years and 4 months" to attend Otto's wedding. But a month later he made a quick trip to Petersburg by ship, despite a sore throat, and in August sailed again for Königsberg and from there went on to Augsburg (Bavaria) to take up his restless way of life. He apparently did not return home during his mother's few remaining years. He continued to write lively, contented letters from his travels. In March 1865 he was invited to play a starring role in Graz, Austria, and wrote that he was considering remaining there, but later that year writes happily from Bamberg.

Emilie wrote at this time, in regard to hearing from Fritz: "What a wondrous thing a mother's heart is, how happy his contentment makes me, and yet I do not approve of the way of life he has achieved for himself and which he is following with determination and consistent purpose. Yes, my children must admit that I definitely have not interfered in their lives, with all three going self-chosen ways. Therefore they are satisfied, for they love the work, creativity and independence. Angels they are not! But they have much diligence, serious purpose and self-respect, and — the blessing of their parents —."

The final entry from 1866, the year of her death, notes that Fritz is in Berlin, "studying art." Carl Bernhardi was in medical school in Berlin at that time, earning his degree in 1867. If Protar studied any medicine, it could only have been at this time, and for a short period.

Though he had earned considerable esteem in heroic roles on various stages, Friedrich Parrot also engaged in writing, chiefly about theater reform in Germany, and he turned more and more to directing. After holding positions as stage manager in Bern (Switzerland) and Oldenburg (northwest Germany), he accepted the position of artistic director at the Riga (Latvia) city theater in 1869, which he held with considerable distinction; in fact, his five years are considered among the best in the history of the Riga stage. He appeared rarely as an actor during this time, but when he did it was "not without success" (Riga Theater and Musical Artists Lexicon, 1890). Outside of the theater, he acted as declamatory speaker for large musical works with accompanying text, such as "Ruins of Athens" and "Manfred." Some time before he came to Riga he had married Anna Strutz, who was reportedly from Dresden. Nothing more is known about her. There were no children.

On the morning of March 28, 1874, Friedrich Parrot left home to attend a rehearsal at the theater, but he did not arrive there and was never seen in Riga again. His wife waited a while for his return or some word from him, but when nothing came, went back to Germany. An enterprising Parrot relative visited her many years later in Frankfurt, but she was not interested in further contact with this family, which is understandable under the circumstances. Considering the nature of his sudden disappearance, he could not have taken much more than his shaving kit with him. It is presumed by the family that he went down to the docks and boarded a ship going west. There was no evidence or even hint of any irregularity — financial, professional or marital — connected with Parrot's sudden disappearance, and it remains a mystery to this day. The only explanation may be political, considering the increasing loss of autonomy and freedom by the Baltic provinces under a "russianization" program, and the character of Protar, being a Parrot and therefore a champion of liberty, and being in a position to act effectively, as in the choice of dramas to be staged. However, even the Riga Theater Lexicon, published only 16 years later and including some very candid comments on other personalities, gives no hint of controversy, only a cryptic "Reportedly living in America."

America 1874-1893

It has not been established when and from where Protar came to America and where he was in the interim; this will probably never be known. Passenger lists of ships are available, but they give the names for the First Class bookings only, which would be very unlikely to include Protar because he could not travel in such style and not anonymously, which he certainly would have preferred. At any rate, he was in New York by December 1874, as verified by notations in a book he bought there, including the first appearance of his American pseudonym "F. Protar." He bought another book early in 1875 in St. Louis, but had not gone to that city to perform; it is significant that his friend Dr. Carl Bernhardi had first gone to St. Louis to stay with friends when he emigrated to America in 1869, but had remained only briefly before moving to Rock Island, Illinois. Protar is known to have given solo dramatic readings during the 1875-6 theater season in New York and February-March 1876 in St. Louis, which were well received by audiences and critics. So far he has not been found in any other cities on the German-language theater circuit.

It has not been possible to link Protar with Fanny Janauscheck and her theater company, whose appearances are well documented and took place on the English-speaking stage from at least 1873, in New York and other cultural centers at least as far west as the Mississippi. The German theater in America was on the decline in the 1870's as the result of a post-Civil War wave of puritanism that frowned on the Sunday performances necessitated by local amateur companies — only the leading performers were professionals and they toured in their roles — and the use of beer halls that doubled as theaters. Many German stages went dark during this time. Protar's serious dramatic readings were more acceptable in this atmosphere. The sources in this area have not been exhausted but the most useful ones are difficult to get to and it may be some time before they can be studied.

It also has not been possible to verify Protar's association with any German-language newspapers or magazines except the one in Rock Island. The editors of those he is supposed to have worked on in this capacity are listed in reference works, but he is not among them under any name. He may well have contributed to such periodicals, but probably anonymously, especially in the case of reviews and critiques which were customarily published without by-lines in those days.

It has now been established that, contrary to earlier conjectures, there was some contact between Protar and his family in Europe while he was in Rock Island (1882-93). The link was discovered by deduction based on a photograph of his brother Otto (1836-93), one copy of which is in the Beaver Island museum and two among the family papers in Munich. These prints, enlarged from a calling card, were made by a Rock Island photographer who was in business at the location stamped at the bottom of the photographs only from 1888 to 93. The calling card, in turn, was taken from a larger family photograph made late in Otto's life. It must have been sent to Protar while he was in Rock Island, before Otto's death (May 31, 1893) or, what seems less likely, shortly after Otto's death, to Beaver Island. This was most probably done by his sister-in-law Helene, Moritz's widow, who was living in Livonia, rather than by Otto's family, who were living in Orel, Russia at that time. This may explain why the family now has two prints: one that was sent to them, and Helene's, that came to them after her death (1933).

Beaver Island 1893-1925

The relatives Protar corresponded with in the 1920's were Otto's daughters, who had grown up in Orel (Oryol), Russia, where he had been head of the telegraph department of the Oryol-Vitebsk railroad, and who had moved with their families to Riga in the then independent republic of Latvia in 1919. These three nieces and their brothers Moritz and Egmont were Protar's heirs. According to estate records (Probate Court, Charlevoix), the heirs in Latvia and Moritz, in Yugoslavia, received their shares of the estate (in 1928), but Egmont, living in Moscow where the United States had no diplomatic or consular representation at that time, did not, and it reverted to the "missing legatee" fund of the county.

The pictures Protar sent to his nieces and especially the notes he wrote on the back reveal important new information about his life on the island. Taken in 1896-8 but not developed and printed until around 1920, these include two views of the house (front in summer, back in winter), Harry the horse, goats on the woodpile, and the well-known picture of Protar on his front porch (Vol. I, pg. 50). They show that

the back and the sides, but not the front of the house were covered with siding. Protar wrote that he added the front "verandah" and back porch, dug the Michigan cellar, made the "big window" (apparently moving the front door), and some "arty" furniture for the front porch in 1893; a crude small table and some benches still existing at the house may be identified with these. "Well, it's all still holding up," he wrote in 1923; it still is in 1980, except the back porch, which collapsed about ten years ago. It is hoped that it can be rebuilt, to restore the house to be the way Protar knew it.

Protar mentioned that he kept three goats for milk and butter "and I made good butter. Tempi passati . . ." he added, saying they had died in the hard winter of 1903-4 and he had not replaced them. Besides much heavy snow, which was not unusual, his weather diaries note many days with temperatures below zero, December through February 1903-4, with 32° below in mid-February. He added that he gave his last horse away in 1916 (Harry had died in 1913). He virtually gave up his garden during the later years, saying that it was only "hand-size" in 1923. "Peas, beans and squash have survived, everything else has been swallowed up by weeds. Do you have squash? A noble fruit, like pumpkin but more refined." He grew some flowers near the front of the house. The vines on the porch "are common hops, easy to get rid of, greening in summer only and beautiful." The elm tree to the west of the house, that has since grown up and died, is seen as a sapling in the pictures. Irene Treboux, Protar's great-niece, remembers heavy wool caps (probably from Sears) that he sent to her and her brother when they were children in Riga, and a dollar bill for each, which they treasured.

In August 1923 he wrote: "These lines must also be written with interruptions. The eyes, or the eye: old; otherwise me: young." Further, at about this time: "The garden is small and poor, weeds abundant, which I can no longer master. There is plenty to do in the old house, chicken coop, etc. Thus I await the 'final day' equanimity, which will bring me not obliteration but only a transformation. For the person who lives a spiritual life there is no death."

It is characteristic of Protar that he did not mention his doctoring activities on the island to his family, although he was busy with them to the end. The newer findings entirely support the character of Protar as brought out by the earlier material, and shed some light on its development and source, his heritage as a moral, thinking, articulate man and one who traveled to find himself in the world. The one flaw that has been revealed in his "saintliness" is the fact that he walked out on his wife, but he apparently had compelling reasons. It is interesting to note the importance of the month of March throughout his life as a time of new beginnings: his birth, of course, and then leaving for Dresden, leaving Riga, leaving Rock Island for Beaver Island, and finally, taking leave of his life, the long-awaited transformation, not an end for him but another new beginning.

St. James Post Office.

BEAVER TALES

by Phil Gregg

St. James, Michigan 49782

The year of 1904 wasn't recorded in the history of the Islands as a year of any particular importance but it was a time of much activity. The commercial fishing was going great guns and the saw mills, too, were centers of activities. Business on the Island was good and showed promise of becoming better. Money was being made and more ways of spending it were being planned.

Johnny Green was an Island businessman who dealt in real estate, a little politics and also farmed. He owned a piece of property on the Main Street of the harbor area and had decided to put it to use. A two story building was to be built and William Boyle was to operate it as a saloon.

John Green.

The building was started on one Saturday and by the following Saturday it was complete. The building was soon referred to as the "Silver Top Saloon," for Silver Top was the brand of beer that was dispensed in the establishment. The upstairs living quarters were used for many a card game and for the convenience of the players a scuttle was cut through the floor so refreshments could be passed up from the bar below via a box with a rope attached. The Silver Top flourished until Prohibition put an end to its happy times.

The building was then used as a tailor shop for a few short years. Frank Nackerman recalls that this was where he was fitted for his first pair of long pants. After the tailor moved out of the building, it stood vacant and neglected for some time. The downstairs doors were either removed or fell off. Cattle, dogs and cats passed in and out; needless to say, the building was soon a shambles.

In 1923 the building was pressed into service again and has ever since been the Post Office of St. James. Willie Boyle, who had been the proprietor of the Silver Top, was appointed Postmaster and served until 1939, when Frank Nackerman succeeded him. Frank Nackerman served until 1973, when Mary C. Minor was appointed Postmaster.

The following people served as Rural Mail Carriers for the Post Office.

Carriers	**Sub-Carriers**
George Williams	
Tom "Lou" Gallagher	
Everrett Cole	Earl Malloy
James W. Gallagher, Jr.	
Frank O'Donnell	
Edward "Bub" Burke	Edward "Gene" Burke
William "Willie" Schmidt	Edward "Gene" Burke
Rose Connaghan	

It might be interesting at this time to give a complete history of the Post Office Department on the Beavers, as compiled by David Colyer of Birmingham, Michigan.

Beaver Island, Michilimackinac County, Michigan

Established as a Fourth Class Office on January 13, 1849
Re-located in Emmet County on January 29, 1853
Name changed to Saint James on October 10, 1854
Re-located in Manitou County on February 12, 1855

Discontinued on March 6, 1868
Re-established as a Fourth Class Office on December 19, 1868
Re-located in Charlevoix County on May 27, 1895
Domestic Money Order Business established
Relegated from a Fourth Class Office to a Third Class Office on July 1, 1945

Postmasters	Dates of Appointment
Charles R. Wright	January 13, 1849
Peter McKinley	September 19, 1851
Dennis Chidester	February 24, 1852
Edward Chidester	June 27, 1856
Bradner Curran	August 1, 1856
James M. McKinley	September 19, 1857
D. D. Lobdell	May 1861
John O. Mala	December 12, 1862
Charles R. Wright	November 20, 1863 December 19, 1868
George Miller	November 4, 1881
James Gibson	December 6, 1881
Ida Gibson	April 2, 1903
Michael J. McCann	July 3, 1903
William W. Boyle	January 23, 1923 (Acting) March 9, 1923 (Commissioned)
Frank J. Nackerman	August 15, 1939 (Acting) February 20, 1940 (Commissioned)
Mary C. Minor	July 1, 1973 (Official in Charge) November 10, 1973 (Commissioned)

Fox Islands, Leelanau County, Michigan

Established as a Fourth Class Office on March 11, 1906
Discontinued on November 30, 1917
Postmaster — John O. Plank Appointed March 11, 1906

High Island, Charlevoix County, Michigan

Established as a Fourth Class Office on June 5, 1913
Discontinued on June 15, 1914
Postmaster — Gladys Hill Appointed June 5, 1913

Nomand, Charlevoix County, Michigan

Established as a Fourth Class Office on March 10, 1917
Discontinued on October 31, 1933
Postmaster — Julia Cole Appointed March 10, 1917

Success, Charlevoix County, Michigan (Garden Island)

Established as a Fourth Class Office on June 3, 1913
Discontinued on February 15, 1914
Postmaster — Richard W. Ballensinger
Appointed June 3, 1913

In order to keep our readers up to date on our present postal service, we will pass on this additional information.

The Zip Code for St. James is *49782.*

In addressing mail to the Island, it should be addressed *St. James, Michigan,* not Beaver Island, Michigan. Putting Beaver Island on the address is permissible as long as St. James, Michigan is below it.

* * *

From the homestead cabin of Egedius and Christina Larsen, on Garden Island, the healthy cries of a new born son could be heard. Arthur Larsen came into the world on the 30th day of January in 1898.

In those days, Garden Island was a busy little community all of its own, with a school, church, post office, store and dottd with homesteads and net sheds of commercial fishermen.

The Larsens' closest neighbor, Uncle Pete Jensen, had several sons, who with Art, kept the two households rollicking with activity.

Fishing was big business and Art's father was doing well, bringing in huge catches of Lake Trout and Whitefish. In 1906, the family moved to Beaver Island and built a fine new home on the harbor of St. James.

Being born to the sea, young Art found the feel of a deck under his feet much to his liking, so handling his father's boat almost went hand in hand with learning to walk and at an early age, Art began fishing with his father. Each day he learned just a little more. Each trip out was a new lesson, whether it was learning the habits of fish, the way the boat handled in a particular sea or what the clouds predicted the weather to be the next day. As the years rolled by it seemed as though Art would never stop growing, for soon he was head and shoulders over most men and was dubbed "Big Art." Although of a gentle nature, one of his cousins relates "nobody gave 'Big Art' any sass."

Art's mother died in 1913, and he continued to work with his father until his father's death in 1923. Both parents are buried on Garden Island.

His bachelor days ended after meeting Sybil Tilley of Beaver Island, and Art and Sybil were married in 1918.

November 11, 1940 was the day of the Armistice Day Storm that stirred the lakes into a fury more severe than ever had been recorded and is a date that is vivid in the memories of those who make their living from the waters of the Great Lakes. For many it ended their wage earning days. Many vessels and lives were lost throughout the lakes but this tale relates only to Beaver Island.

With winter closing in, Island fishermen were anxious to wind up the season on a successful note. Long before daylight engines were started and one by one the Island's small fleet of fish tugs headed out of the harbor to lift and re-set their nets.

"Big Art" Larsen's tug "Estonia" headed for his sets off Gull Island. There was no wind but heavy swells were rolling up the lake from the south. Starting to lift his nets, "Big Art"

The "Estonia" during building.

was amazed at the amount of trout in his nets. With only four boxes of nets lifted and re-set, his boat was heavily loaded with fish and there were still 36 boxes out. With an eye out for the weather, Art could see a storm bearing down from the southwest. It was not a dark cloud but almost like a fog bank moving up the lake. He decided to head for home. The skippers on the other tugs saw the ominous signs of the impending storm also and headed for St. James.

Arriving at his dock, Art tied his boat and he and the crew went about cleaning their catch. The other tugs were coming in; Frank Left's, Andy Gallagher's, Roland McCann's, and coming in right with the storm was "Shing" Martin's tug the "Evelyn." No time to tie up, Shing headed his boat into the beach and ran her right up on the sand. The mail boat, "Mary Margaret," tied up at the old Booth Fish Dock, was straining at her lines and was beginning to tear up the dock. The alerted crew fired her up and cast off to "run out" the storm in the harbor. The other tugs, except the "Estonia," did the same for the wind was blowing with such force lines couldn't hold them.

As each hour passed, the wind increased in velocity and the fight to save the boats became a true test of stamina and seamanship. The water level in the harbor was rising rapidly and soon seas were breaking over the top of the various docks. The lines that held the "Estonia" to her dock parted as the level rose and just as though she had a phantom pilot at her wheel, she backed out away from the dock and maneuvered her way to the northeastern end of the harbor and put her bow gently on the beach, between two ancient shipwrecks. Anxious observers, peering from their windows, were certain "Big Art" had for some reason untied his boat and put her on the beach. Darkness had fallen and the drama in the harbor continued. Lighted boats struggled against the storm.

With the light of day the storm had subsided and water level was nearly normal. Debris was piled high on the beaches. From all appearances and considering the force of the storm, it looked as though Beaver Island had survived quite well. Shing's and Art's boats were high and dry but undamaged. Docks were in disarray but easily repairable. The real damage was done out in the lake where these people made their living. The many boxes of nets were gone, torn from their sets and swept away. Art had forty boxes out, each box consisting of four nets, and there were ten boxes to a gang. A gang of nets meant a loss of nearly $1,200.00. Shing Martin had lost even more and, of course, the other fishermen had similar losses. This was the turning point in the Great Lakes fishing business, as many had lost expensive equipment and the trout were getting scarce.

The "Estonia" on ice after the storm.

Though fishing was in his blood, Art turned to sailing as a diesel engineer on boats plying the northern waters, such as the "White Swan" and the "North Shore." Being an ingenious mechanic, it has been said "Big Art" could make a diesel run on soft coal if he had to.

His colorful life came to an end June 3rd, 1964, after a year's illness. Arthur "Big Art" Larsen, was laid to rest in St. James Township Cemetery on the 6th of June.

* * *

This story was told to Phil Gregg by his friend and neighbor, William "Willie" Belfy in 1964:

My first look at Beaver Island was in 1908 while serving as crew on my step-father's two masted schooner, the "Waleska." Quite often we would pull into St. James to pick up a cargo of lumber, cedar posts, barrel staves or railroad ties. The harbor

The "Waleska."

was bustling with activity and was always an enjoyable place to stop, especially for a young lad of eighteen.

Working on the Lakes under sail was a far cry from to-day's mode of lake travel. Men under sail not only had to be expert sailors, but had to be weather prophets, buyers, and salesmen as well. My step-father, Herman Ludwig, was one of these men, calm under the adverse conditions which were common in late fall.

December of 1910, we had left St. James with a cargo of barrel staves, headed for Ludington. We encountered gale winds off the Manitou Islands with driven snow so thick one end of the vessel couldn't be seen from the other. The 'Old Man,' who could judge the speed of the vessel by looking at the wake and plotting a course from his last visible sight before the gale hit, headed for Frankfort. After running his estimated time he headed her toward the piers. With the sails snapping in the gale and snow stinging our faces, we listened intently for the horn on the pier. The 'Old Man' shouted, "If we miss her, we'll walk ashore, boys." When we first heard the horn we could then make out the piers — on either side! We had made it right on the button.

In 1912 I married Alice Johnston, of Beaver Island, and went to work for the Beaver Island Lumber Company, piling lumber on the dock and loading vessels. Between two hundred and two hundred and fifty thousand feet of lumber a week were being shipped from St. James. The train that ran to the head of the Island had to make at least two trips a day, with eight cars, in order to keep the mill busy. Often times they had to head the train up with two engines and make four trips a day. During the summer months tugs would tow booms of logs up from the head of the Island to feed the mill. At this rate it was no wonder the timber was taken off so fast.

From 1915 to 1935, my brother Erwin, a hired hand and I commercial fished with the tug "Silver Star." These were the days when catches were measured in tons, not pounds.

I remember one December during this period when we were caught off the Fox Islands in foul weather for ten days. It was freezing too hard to lift our nets, so it was just a matter of waiting it out and shifting anchor from one lee to the other around the Island.

Between 1935 and 1937 I served with Captain Allers on the Conservation Patrol Boat No. 1. We patrolled all of Michigan waters, from Michigan City to Isle Royale, in Lake Superior.

The "Silver Star."

I went back to commercial fishing again from 1937 to 1940, with Ronald McCann on the tug "Venus II." Fishing was falling off. The era of the big catches was over and a war was coming on.

I left the Island to work at the Fisher Boat Works, in Detroit, where I helped build 110 foot Sub-chasers for the Navy.

I returned to the Island and went back to fishing with Roland McCann until 1950. By this time the fishing had practically stopped and the "Venus II" was one of the last tugs to fish gill nets from Beaver Island.

The bones of the first boat I served on, the "Waleska," are now bleaching on the beach, just east of Charlie Martin's dock. The tugs are gone, the mill is quiet. The Island is in a new era and I now serve a few hours a day as bartender at my daughter's tavern, the "Beachcomber" in St. James.

(A model of the "Waleska" will be on display at the Beaver Island Marine Museum.)

* * *

1921

Times were good and Peter Boyle decided this was the opportune time to expand his operation by gracing his farm with a brand new barn. Competition among the farmers of those days was keen and certainly Peter was not to be outdone. It was his desire to build a barn bigger then any other on the Island. Tommy 'Lew' Gallagher, Peter's uncle, was contracted to do the job with the understanding the barn be built, "a foot wider and a foot longer than any other barn on the Island, and build her just as high as you can get her." The barn was built and a fine one it was, with high double doors in the center of both sides so large hay wagons could easily unload and not have to back up, but go right on through.

Peter had been highly impressed, the last time he had been on the mainland, by a new fangled contraption called a horseless carriage. Detroit's addition to modern man's whims had made its way to the north country, and Peter wasn't going to be denied the opportunity of owning one of these machines.

A few weeks later a brand new 1921 Model T Ford was rolled off the deck of the mail boat, where its proud owner went about getting the darned thing started. Most everyone in the gathering crowd had a suggestion or two but in spite of this, Peter got it started and away he went in a cloud of dust, with barking dogs and bolting horses adding to the excitement. Turning in the gate at his farm, Peter's father came out of the house to find out just what in tarnation all the noise was. Peter shouted to the old man to open the barn door while he drove in circles in the yard. Peter then headed his shiny new steed for the barn. Regardless of how loud he shouted "Whoa!" the machine wouldn't slow down a bit, and it didn't find much resistance when it had gone through the barn and came to the doors on the other side. With a splintering crash, the monster bolted through the door and continued on its way until a sturdy fence post halted its forward progress. Learning to stop the car seemed to be the big problem to solve and this was soon accomplished by stacking several bales of hay against a large maple tree in the yard.

One day Peter decided it was time to replenish some of his personal supplies and after starting his iron horse, headed for town. With no padded "stopping trees" near the store, he felt fortunate when he saw Nels LaFreniere sitting on the steps of his store. Shouting to Nels his order for Peerless tobacco and other sundry items, he chugged on about the harbor to the lighthouse, thus giving Nels time to fill the order. Upon Peter's return, Nels was waiting and tossed the order in the car as it passed slowly by the store. Returning to the farm, Peter found his "stopping tree" proved a huge success and this saved considerable wear and tear on the wonderful new barn.

BAR
MODEL
Beaver Island - St. James Mi

The Brother Domnan Memorial on the Maloney Estate at
St. James, Beaver Island, Mich.

Sisters Home, Holy Cross Church, Beaver Island, St. James, Mich.

First House built by white man on Beaver Island. 1842. St. James, Mich.